Cliff's Nodes

Cliff's Nodes

Editorials from *The Physics Teacher*

Clifford Swartz

The Johns Hopkins University Press

Baltimore

The Johns Hopkins University Press
2715 North Charles Street
Baltimore, Maryland 21218-4363
www.press.jhu.edu

Library of Congress Cataloging-in-Publication Data
Swartz, Clifford E.
Cliff's nodes : editorials from The physics teacher / Clifford Swartz
p. cm.
Includes bibliographical references and index.
ISBN 0-8018-8306-7 (acid-free paper)—ISBN 0-8018-8307-5 (pbk. : acid-free paper)
1. Physics—Study and teaching. I. Physics teacher. II. Title.
QC30.S93 2006
530'.07'12—dc22   2005052000

A catalog record for this book is available from the British Library.

Contents

Contents

Preface

What a bully pulpit! For twenty-nine years I was privileged to be the editor of *The Physics Teacher*. During those years we developed the magazine from being a missionary effort by physicists for high school teachers to being a journal for the teacher of introductory physics at any level. Of the twelve thousand or so subscribers, about two-thirds teach in colleges and fifteen hundred are outside the United States. That leaves only about four thousand high school teachers in the United States who read the journal, but surprisingly, that's all the full-time high school physics teachers in the country!

During those years I wrote editorials almost every month. The editorials were designed to cheer, chide, and celebrate. Many were folded around tutorials. Not everyone agreed with my biases, but teachers read the editorials and let me know their reactions. Here's a collection of my favorites, with samples of celebration and samples of picky criticism. As for the former, rejoice with me that I think that teaching is a noble profession. As for the latter, enjoy the twist of the words. I wasn't only kidding.

And God said:

$$\nabla \bullet \mathbf{E} = \frac{\rho}{\varepsilon_0}$$

$$\nabla \bullet \mathbf{B} = 0$$

$$\nabla \times \mathbf{E} = \frac{\partial \mathbf{B}}{\partial t}$$

$$c^2 \nabla \times \mathbf{B} = \frac{j}{\varepsilon_0} = \frac{\partial \mathbf{E}}{\partial t}$$

—and there was light.

Part I CELEBRATION

We begin with the celebration of physics and of teaching. The teacher of physics can have a good and rewarding life. There are friendships to be made with fellow teachers in the local school and nationwide through the American Association of Physics Teachers. There is creative work to be done in designing new apparatus and by publishing in journals. There are young people to influence and to keep us young. The subject matter is always changing and forever coming up with exciting surprises. In what other subject can you link everyday phenomena with the nature of atoms? In our classes students can learn that we are truly made of stardust, and they can travel with us as we go from Newton's laws in the classroom to the beginning of time in the new universe.

> Outside of physics, it's all stamp collecting.
> —Ernest Rutherford

> I often say that when you can measure what you are speaking about and express it in numbers you know something about it, but when you cannot express it in numbers, your knowledge is of a meager and unsatisfactory kind; it may be the beginning of knowledge but you have scarcely in your thoughts advanced to the stage of Science, whatever the matter may be.
> —Lord Kelvin

recently had occasion to read again Galileo's *Dialogue on Two New Sciences*. Although it's hard to read Newton's *Principia* with its geometrical proofs, it's a delight to follow the banter and brilliant reasoning of the *Dialogues*. What a marvelous time to have been a scientist! Only three hundred and sixty-five years ago our whole world view was being overthrown by a few men, armed with their brains and some new optical devices. I particularly like Galileo's analysis of falling objects. He says that he has actually dropped balls of different weights from a tower (though not in Pisa). He also points out that the heavier one will hit the ground first, but only by an arm's span! For an arm's span, will you adhere to Aristotle who would have the balls separated by half the height of the tower? I also like Galileo's careful experiment to determine the speed of light. He stationed an assistant on a distant hilltop with a lantern. When the assistant saw Galileo blink his lantern, he blinked back. Of course the method did not produce a delay longer than the reaction time, but Galileo had measured that. He realized the limits of his apparatus and method, and says that he is only producing a lower limit for the speed of light.

Even if we can't now easily read the *Principia*, that age must have been a marvelous time to have been a scientist. Three hundred and thirteen years ago, Newton laid out the mathematical structure that explains the motion of things on Earth and in the heavens. Of course the lion's paw was imprinted on many other topics in physics. He divided the sunlight into colored rays and combined them again. He made and analyzed optical instruments, including the reflecting telescope. It seemed for a while that the foundations had been laid for the final understanding of our universe. Only the details remained, and they were worked out by Euler, Lagrange, and Laplace in the century that followed.

Two hundred years ago, Franklin grounded the lightning. In the next few decades, Oersted, Faraday, Henry, and others discovered that magnetism and electricity are manifestations of the same phenomenon. After each discovery, inventions followed, turning science into technical revolutions. In retrospect, the discoveries seem so simple. There is a legend, for instance, that Ampère kept his ammeters in a room separate from the magnets, and so did not see the current pulses as he changed the magnetic field configurations. It was Faraday who realized, six years later, that there must be relative motion of field and conductors to produce current. Wouldn't it have been marvelous to have been a scientist in those years?

Only a little over a hundred years ago, science was almost finished. That's what some wise men thought. All that remained to be discovered, among other things, were electromagnetic radiation, radioactivity, x-rays, and relativity. There were also several embarrassing anomalies concerning spectra from solid objects, and a complete mystery about the regularities of atomic line spectra. Those were the years to have been a physicist! New discoveries, new elements, and new theories were coming along each year, and many of them would be named after their discoverers.

How could any era in physics compare with the 1920s? With Einstein and Bohr and Pauli and Dirac and Schroedinger and Heisenberg? It was like the publication of Newton's *Principia.* One strange new way of considering nature appeared, and the walls of problems came tumbling down. The structure of atoms was in principle solved. Those events of the '20s and '30s are legendary, the more so because many of us now living knew those people and some of our older colleagues were participants in the quantum revolution.

And then there's today. In the 1960s we started the framework to explain all the subatomic particles. The gaps are all filled in now. In 1999 you can buy a pocket calculator that can tell your position on Earth to within a few meters. Its calibration depends on both the special and general theory of relativity, but you can read it just by pressing a button. Since the 1920s we have known that the universe is expanding. It doesn't seem to affect everyday life. Is this a good time to be studying physics?

It sure is! Actually, the mysteries about the subatomic particles are deeper and more profound than ever. Decade by decade we have learned new things that we did not know we do not know. Why do the particles have their particular masses? Does the neutrino have nonzero mass, and if so does this help explain the missing mass of galaxies and our universe? The sciences of the very large and the very small have always been allied. Witness the role of spectra in revealing the nature of atoms and of stars. In our own *aetas mirabilis* we are after bigger game—the neutrinos play a vital role in the birth and death of stars and perhaps of the universe itself. Every month our telescopes, on mountaintops and in space, solve old problems and reveal new mysteries. On our tabletops at home, our circuits flip polarity in nanoseconds. In the laboratory our lasers condense beams to the width of femtoseconds, allowing us to monitor individual chemical and biological reaction steps. The age of science may not be in its infancy, but it's not far into adolescence. What a marvelous time to be studying physics!

April 1999

Under the Influence

ast May we had an article in *The Physics Teacher* showing that college physics students did not benefit from having taken high-school physics. The analysis was based on a large sample of students and was done with great statistical care. Thus it led to a puzzling, if not discouraging, conclusion. Is there no point in taking physics in high school if you're going to take it in college? Of course, as one Letter to the Editor asked, is there proof that it is useful to take any high-school course that is going to be repeated in college? Furthermore, from what many college teachers have discovered by administering the Force Concept Inventory, the college physics course doesn't seem to improve student comprehension either. It's a lose-lose situation.

But there is good news. When Steve Chu won the Nobel prize, he was asked why he had become interested in physics. On the nationwide Lehrer Report, Dr. Chu gave credit to his high-school physics teacher—Tom Miner. As most of our readers know, Tom was our associate editor for many years. He was a true gentleman and a great teacher. Several generations of students in Garden City on Long Island came under his influence. I doubt that any test could determine how many physics facts his students learned during their year in Tom's class, but I have heard from many of those students that Tom had influenced their lives. Influence like that is hard to define. It's a matter of moral stature, intellectual discipline, and friendly caring.

I thought of these things again when Max Dresden died. Max was a great teacher with a very different personality from Tom's. Max was physically vivacious and intellectually confrontational. If he hadn't seen you for awhile and met you in the hall, he would demand, "So, what have you learned recently?" Max never turned down an opportunity to speak to high-school students or teachers. Even though he was very sick, he gave the Klopsteg lecture at the 1997 summer meeting of American Association of Physics Teachers (AAPT). Of course, most of Max's teaching was at the graduate level of theoretical physics. Over 60 graduate students got their Ph.D.'s under his guidance. Nevertheless, he took his turn teaching the undergraduate courses, including the introductory course where he was a great success.

I remember the Ph.D. final exam of one of Max's students. Max pointed his finger at the nervous candidate and asked, "Who is your teacher?" Flabbergasted, the student finally blurted out, "Why, you are." "Ah," said Max, "but who was my teacher?" "Uhlenbeck?" the student answered. "Yes, but who was Uhlenbeck's teacher?" "Sommerfeld?" the student guessed. "Correct," said Max, "but who was Sommerfeld's teacher?" "Boltzmann," declared the triumphant student. "You see," said Max, "you are in the line of a great tradition."

I flew to Ann Arbor for Dick Crane's 90th birthday party. When you are 90 years old, your students have retired, or taken the other way out, and their students are checking their retirement policies. At the symposium, we heard about Dick's physics research in the '30s and '40s, including his exquisite measurement of the electron g-factor. There were talks about his work in biophysics and in administration—department chairman, president of AAPT, chairman of the board of American Institute of Physics, and many others. We toured his current interest, the establishment of a superb science museum in Ann Arbor. Throughout all these activities there is an underlying theme. Dick Crane is a teacher. We have witnessed this in *The Physics Teacher* with Dick's columns on How Things Work. Not only does Dick figure out the mechanics of common objects, he explains them in ways that we can understand. Dick's influence has spread far beyond Ann Arbor and has affected college teaching throughout the nation.

Our prime role as teachers is not to teach facts, important though they are. After the students leave us, they will soon forget the facts. Instead, for better or worse, teachers are models of how to think, of how to solve problems, of how to help others learn. Our influence affects the generation we teach, and through them the generations beyond. As Max said, we are in the line of a great tradition.

December 1997

Grading the Teacher

everal fads ago there was a movement to grade teachers in terms of their competency—competency-based testing. Everyone knows that there are good teachers and there are bad teachers. The trouble is, it's hard to define the categories. It's like the Supreme Court justice who couldn't define pornography, but knew it when he saw it. In New York State, prospective teachers must take tests in both pedagogy and subject material. That seems reasonable. There ought to be some minimum standards, so I thought that I would try my hand at setting up such requirements.

For pedagogy, I think that a physics teacher should be able to describe some main ideas of educational practice in the twentieth century. What did John Dewey advocate? What were Piaget's contributions? What was novel and what changes were made by the Physical Science Study Committee (PSSC) course?

For science knowledge, how can we measure atomic sizes and masses? What are the relative wavelengths and frequencies (and how can we measure them) for key regions of the electromagnetic spectrum? Perhaps the science requirement could be specified in terms of getting a passing grade in a course one level beyond that to be taught.

Given an afternoon, a pot of coffee, and some congenial colleagues, I think that I could make up a reasonable threshold exam for prospective physics teachers. The difficulty of the test would be determined by what I could answer successfully if forced to take the thing. I have a nagging feeling, however, that the highest scorers on such a test might turn out to be the poorest teachers.

Long ago I ran a workshop for the new PSSC course. Out of about 50 physics teachers (or 50 teachers who occasionally taught physics), only six had taken more than one college physics course. I remember two of these in particular because they were such complete opposites. One was shy, rather inarticulate with adults, and a little embarrassed about his lack of knowledge of physics. The other was flamboyant, always on stage whether with students or their parents. He didn't have much preparation in physics either, but he exploited that fact. If a student asked a hard question, this teacher's most common response was ''I don't know. I've never thought of that before. How will we find out?" And the teacher made sure that he and the class followed through. The shy teacher couldn't have pulled that off. He knew he wasn't a good lecturer and so he divided the class into small groups to tackle problems together, while he wandered from group to group helping with the problems or the experiments. I visited these classrooms often and knew good teaching when I saw it. But I'm afraid that neither teacher would have done well on my competency-based exam.

In graduate school days I took two courses in advanced quantum mechanics. The first was given by a superb lecturer. He arrived for class each day with notes carefully prepared. At the end of the hour you could see the whole lecture neatly arranged on the board. Where the textbook might say, "It can be shown that . . .," our professor would show it for us. It was a joy to watch how smoothly he fit everything together. No one missed any of those classes. The next year our lecturer was very busy with research that was novel and exciting. He would show up for class a little bedraggled and ask where we had left off last time. Sometimes that made no difference because he would rather tell us about his latest results and problems. That was a tough class. We had to do all the learning ourselves. At the time we thought that the first man was the better teacher. But we learned more from the second one. Neither one had ever heard of John Dewey or Jean Piaget.

In undergraduate days I took a physical chemistry course under a man who could easily have won the vote as worst teacher in the school. He sort of mumbled at the board and erased with his left hand what he had just written with his right. I got permission from him to take a math course at the same time and just show up for exams. My friends claimed that I had an unfair advantage. However, I stayed on at the same university for graduate school and came to know this chemist as a teacher of graduate students. He was a great teacher—brilliant, kind, inspiring. His disciples held prime posts in chemistry departments around the country. He just never should have been forced to give large lectures. A "good" teacher may be good for some students, but not others. Teaching is not a process of administering knowledge; the process is one of interaction.

So how do you judge the competency of a teacher? Do you ask for a sample lecture? Imagine listening to Niels Bohr stammer in five languages, circling and advancing and mulling over an idea. But it probably was a great idea! Do you ask for performance data of the teacher's students—perhaps improvements on their SAT scores? You would probably find that the teacher of students from high-income neighborhoods did better than those who teach in low-income neighborhoods. You could reward the former, thus making the situation worse.

I wish that there were a simple way to grade (and properly reward) the teacher. The criteria would have to include background knowledge of physics and pedagogy, but would also have to take into account enthusiasm for learning physics and for learning about the learning process itself. There would have to be extra points for enjoying students and knowing how to be friends but not pals. I'm afraid that I don't know how to test for good teaching, but I know it when I see it.

April 2000

S ome years ago Japanese manufacturers revolutionized assembly line procedures with a system known as "just-in-time inventory." Instead of locking up sizable capital in the form of storage bins full of parts to be assembled, the companies arrange the entire supply system to provide each part just as it is needed. Of course that takes precision planning because one delay can wreck the process.

For two summers I was engaged in a project with high-school students that uses just-in-time education. These youngsters studied the feasibility of establishing a habitat for 10,000 people in orbit within the Earth-Moon system. This is an update of the L-5 project, first proposed 15 years ago by Princeton Professor Gerard O'Neill. My summer students had been selected from throughout New York State on the basis of their good school records in science and math. Most had not yet "taken" high-school physics. For the project, however, it's necessary to understand and use all sorts of physics concepts and formulas. Rocket operation and efficiencies must be figured, escape energies and velocities must be calculated, centrifugal forces must be measured. It turns out that people can learn to do these things without our standard buildup of sequential lessons. One of the prime requirements for learning appears to be the "need to know."

In the security business, access to classified (secret) information is not determined solely on your record of good citizenship. There must also be a need to know. This phrase has a different connotation for school learning, but the idea is every bit as important for learning as it is for national security. Without the need to know something, preferably for immediate application, learning becomes tedious and superficial. The students in our project needed to know about centrifugal force because they had to design a rotating structure that would provide normal gravity for 10,000 people. Note that in the usual "course," which we "give" and the students "take," the main need to know about centrifugal force is to be able to plug numbers into the right formula on the exam, or perhaps on a multiple-choice test to differentiate between centrifugal and centripetal.

Wouldn't it be interesting to design a physics course, or better yet a whole school curriculum, consisting of a number of short projects? Some of them could be simulated adventures such as the space habitat. Others could be real-life projects, perhaps analyzing or helping to work on local problems. Such problems abound—from the individual, personal needs of growing up to the environmental or political needs of society.

There is precedent for devising short blocks of learning based on themes. Almost 20 years ago a group at Tallahassee started the Intermediate Science Curriculum Study project to produce such materials for high-school science. The themes of these nonsequential units covered a wide span of human interest and technical needs. The project fell upon hard times when National Science Foundation funds dried up and the idea never received an adequate trial. Another precedent for using real-life projects to create a need to know is the Scout merit badge system. This system has its difficulties with volunteer supervision, but when it works well, it produces long-lasting learning.

Now that we are adults, how do we go about learning anything? Do we simply go to a school and ask to be assigned a course in any subject that the system ordains for us? Of course not. Next week, to meet a deadline, I must write a paper on the motions of fluids. I have a very real and pressing need to know about viscosity and mean free paths and scaling factors. Nobody is offering a course about these particular details, but because I have tackled a lot of similar projects, I know how to look up references and consult knowledgable friends. Probably I learned some of these things about fluids years ago, but then I had no need to know and the learning didn't stick. Now the pressure's on. I'll learn efficiently and thoroughly just what I know I need.

Maybe that's the way we ought to let students learn. Present them with, or let them create, a need to know. Then help them learn whatever they need to know, and just in time.

October 1990

An Answer for Joanna

My name is Joanna Bare, and I am a junior in high school in Manitowoc, Wisconsin. I am interested in a career in physics and/or astronomy. I would like to know the names of the best colleges and universities in these fields.

Important questions seldom have simple answers. Indeed, there may be no answer at all. Even if you choose your college on the basis of sage advice and sound facts, you will still be subject to unknown fates that may be more important than any you can now foresee. A key professor may go on leave, or a visiting professor who can change your life may happen to be there. In choosing a college you are also choosing friends, satisfactions, disappointments, and excitements. About these you can only gamble. (Here's a sobering thought. When my wife was your age she went off to college early. On her first night there she had a randomly selected blind date with a boy who became the father of our six children.) But there are things you should know about colleges that can help the odds.

First, there are colleges and there are universities. The traditional goals of each are different. Universities offer the Ph.D. degree. The faculty in a university have a mandate to do research and to train graduate students. In some particular university, there may be great undergraduate teaching in a particular field, but it's probably not the main interest of the department. It shouldn't be. Some of the most productive and best known physicists in a department may seldom teach undergraduates. Furthermore, some of the undergraduate teaching—labs at least—may be done by graduate students. This situation may not necessarily be bad. Some departments have a tradition of rotating senior professors through the major undergraduate courses and of providing teacher training for the graduate students. (In general, college and university professors get no such training.) Some departments have active physics clubs and encourage undergraduates to join research teams. In general, however, the university atmosphere is high pressure and individualistic. Opportunities abound, but behind doors that will open only if you knock. There will be fascinating talks by the leading figures in physics research. You can go and let the exotic words wash over you. Or you can stay in the dorm and play cards. No one will bother you either way. A university is not the place for people who need hand holding. It's a place for people with the maturity and temperament for intellectually elbowing their way forward.

The prime mandate for a smaller college is to provide good teaching. The faculty may be expected to publish occasionally but usually only as a demonstration that they are keeping up in their field. If research is done it is at a slower pace. Usually there are only a few members of the physics department and no graduate students. Faculty-student contact is expected to be closer. You may have one particular teacher for several courses, and get to know very well the few other physics majors. Good small colleges make an effort not to lose students, socially or academically. There is apt to be a greater feeling of belonging to the institution, on the part of both faculty and students.

Where can you get the best undergraduate physics education—college or university? There are excellent departments and opportunities in both. The choice depends partly on you and your style. But there are questions to ask any college or university you visit. Who teaches the introductory courses? Who teaches the recitations and labs? How many students are in each section? Are any undergraduates helping in research? How active is the physics club? Are members of the faculty active in research, and are they active in their professional organizations? (In the case of physics, do they belong to the American Physical Society and the American Association of Physics Teachers?) When was the instructional lab equipment bought, particularly for the advanced courses?

There's another key thing to know about becoming a physicist. The undergraduate training is only a beginning. You can get good training in either small colleges or large universities. But then you'll want to get into a good graduate school. For that you'll need good course grades, good college board exam scores, *and* good recommendations from your undergraduate teachers. It will help if they know, and are known by, the physicists in the graduate schools. That's another reason why you should find out if the undergraduate teachers are active in their field. When interviewing colleges, ask about their track record in placing their majors in good graduate schools. Be very specific in asking the question—like, where did last year's physics graduates go? (If you're embarrassed to ask these questions, urge your parents to ask.)

Incidentally, you should realize that for creative work in physics, or for advanced physics teaching, it's almost essential to have a Ph.D. Currently, the average time to get the degree, after the A.B. or B.S., is five years. Your parents should know that most graduate students get assistantships that pay the tuition and enable the students to live in ennobling, if Bohemian, poverty.

Everything that I have been saying applies to astronomy as well as physics. Indeed, in many universities and colleges the undergraduate major programs are almost the same. A physics major is well prepared to enter most astronomy graduate programs, though the converse is not necessarily true.

There is another factor that you should take into account in choosing to major in physics. It is a field that pays no attention to who you are or where you come from but demands only creative ability and accomplishment. Don't go into physics unless you really enjoy—and are good at—mathematical and technical work. Regardless of your high-school enthusiasms, keep an open mind in college about other interests. An *education* should literally *lead one out*. Especially for physicists there is a tradition of being well and liberally educated. Your college years should provide you with such opportunities, and this factor should enter into your choice of college.

We need bright, creative people in this profession. Our domain extends from the subnuclear particles to the far reaches of the universe, from the beginning of time to the decay of mesons. There are many routes to enter the profession. If the road seems long, keep in mind that the great insights often come to young people along the way. Shop around for a college or university that has creative physicists and shows interest in training more. Then hurry, and come join us.

May 1984

Auld Lang Syne

was on my way in, a little late, when I saw the old codger leaving. He was wearing the traditional garb—you know, white toga, sandals, slide rule as long as a scythe.

"Hey, Pops," I said, "You're overdoing it. Every New Year's it's the same. You look like a leftover cliche."

"Kid," he said, "Some of my best friends are cliches. Besides, what about yourself? You still have your sash on and already you're looking for a place to plug in your microcomputer."

"Just standard wave-of-the-future," I replied. "Got any advice to give, resolutions to pass along? It's traditional, you know—and no offense, Pops, but the way you look, you don't have much time."

"Yeah," he muttered. "What a year! Downhill all the way. But I'll tell you one thing, they're doing it all wrong. Don't let 'em send you to grade school, Sonny. And if they do, don't let them teach you science. They probably won't, anyway. Pretty much cut it out everywhere. Back to basics. Good thing, too. Most of it was being done wrong. They had the kids memorizing all sorts of names they didn't understand—energy, gravity, molecules, genes, solar system. Had them looking through microscopes at their own eyelashes and watching cartoons. Called it science. Kids didn't understand any of it, of course. Lots of educators talking about Piaget, and then trying to teach quantum mechanics to grade school kids."

"Well, they have to learn it sometime, Pops. What would you teach grade school kids?"

"Why, I'd teach them to use hand tools and how to measure things. I'd teach them how to put their measurements on graphs. By the time they went to junior high they might know how to add, subtract, multiply, divide, and plot all the things they could get their hands on— lengths, weights, times, temperatures, voltages."

"That's ancient, Pops. No need, anymore. You want arithmetic, let the computer do it. You want experiments, simulate them."

"Then there's junior high. More antiscience. Most texts for seventh graders, and eighth graders, and ninth graders claim they're going to teach everything there is to be known on the subject. A lot of the authors know it's wrong, too. Doesn't make any difference. Kids can't learn it. Earth science, they memorize the name 'Coriolis force.' Physical science, they learn Newton's laws. Life sciences, they look at DNA. Flaky little fourteen year olds, and they try to teach them quantum mechanics."

"Well, they have to learn it sometime, Pops. What would you teach junior high kids?"

"Why, I'd teach them to use hand tools and how to measure things. Let them take their bikes apart and put motors back together. Let them pull on pulleys and pry with levers. You learn more science in shop than you do in most science classes. Maybe they could also learn to graph data."

"How about high school, Pops? Things going better in high school?"

"Worse. Young teachers leaving. Old ones getting older. No new ones coming in. Not trained ones, anyway. Doesn't make any difference. It's still all backwards. First they teach modern biology—big chain molecules. After the kids memorize that, they take chemistry and learn what molecules are. They memorize orbital schemes and energy levels. After that, a few of them take physics. Learn what energy is. Don't understand it, of course, because they never learned how to measure anything. Never learned how a pulley works, because all they teach them is quantum mechanics."

"Well, they have to learn it sometime, Pops. At least they all learn calculus in high school these days."

"What calculus? They still can't do algebra. They can't use trig. High school kids aren't old enough for calculus unless they're already good at algebra and analytic geometry, and practically none of them are."

"College, Pops. Surely they're doing it right in college?"

"Sonny, you're too young to smoke that stuff. What they've done in college is to move quantum mechanics down to the third year, or maybe the second. That's because the freshmen already know calculus—and can take a much more advanced introductory course."

"But you said high school seniors can't learn calculus."

"They can't. So most of them drop out of these college physics courses. Or the colleges give remedial courses to teach the kids how to measure things, pull on pulleys, and draw graphs.''

''Well, they have to learn it sometime,'' I said.

He went grumbling away down that long road, occasionally waving his slide rule. I would have laughed and laughed, but it hurt too much.

January 1983

The Uniform of the Day

On 15 October at 0700 my students here at West Point started wearing long-sleeved shirts. So did I. It's part of the uniform of the day. We all wear uniforms, in every profession. For physicists the uniform is usually one of ostentatious informality. Our working clothes are jeans, unpolished shoes, and, of course, no tie. Female physicists tend to be neater, but clearly are not dressed for the front office.

I remember one occasion long ago when the Cosmotron first started up. It was a network news event. Reporters and TV cameramen were setting up their gear in the control room while the staff helped provide power lines and stands. Technicians, engineers, physicists, we were all dressed in our standard working clothes. "All right," said the chief reporter, when everything was ready. "Now send in the scientists." He apparently expected people in white coats.

The informality among physicists is deliberate. Everyone on the team answers to his or her first name. An outsider might well think that the system shows lack of proper respect and would breed anarchy. On the contrary, the system emphasizes enormous respect, but for ideas and technical proficiency, not personages. One time at Brookhaven Lab, Owen Chamberlain, who later won the Nobel prize, was trying to figure out how to run a particular instrument. The technician who was showing him became exasperated. "Look," said the technician, "I will explain it to you so an idiot could understand." Now the moral of the tale is not that such a comment was made, but that everyone, including Chamberlain, thought that the remark was appropriate and admired the phrasing.

There is a legend in our business that when Wolfgang Pauli was about twenty years old, Einstein came to his university to give a colloquium. After the great man had spoken, Pauli got up and commented, "What Professor Einstein has said is not so stupid." Or perhaps, the translation from the German should be "is not quite wrong." Or perhaps the story is apocryphal. It doesn't matter. The point is, it's a legend in our profession which we pass along as a model of the proper respect due authority and custom. Note I'm not talking about disrespect, but proper respect. And for a physicist, that means a challenging, inquiring attitude. We should breed it with our traditions and folklore and daily working customs. We should also breed it in our classrooms.

It's hard to persuade introductory physics students that the laws of physics can and should be challenged. Most students have all they can do just to memorize those hundreds of special formulas. But a physics course that presents our present system as a collection of graven laws perpetuates a fiction and does a disservice to our enterprise. The facts we teach are not complete, the laws are not necessarily in their final form, and we deal most of the time with approximations. It's important that the student knows that this is the situation. It is more important for the student to ask new questions than to learn our answers to old ones.

It's easy to train young people to say "Yes, sir!" In physics, most students will gladly conform to any strange custom that will provide them with correct answers, as determined by the grades. We must teach a harder thing—to ask "why?" Our standard uniform in class, regardless of what we wear, should be a habit of questioning, of challenging, and of finding new ways to view the familiar. Not that the students should get too fresh, however. After all, what we say to them may not be so stupid.

November 1981

hen I was a boy there was a woman's magazine called *True Confessions*. It told shocking tales, for those days, of loss of virtue and wild abandonment of wifely duties. One of my aunts took the magazine. Of course, I was not allowed to read it, but of course I did. Fortunately, there was a moral to every tale.

Now I have something to confess, more shocking for a physics teacher than any temporary lapse of middle class propriety. Close the doors, and gather round.

This past month I have been doing plug-in homework problems— hundreds of them. I did them by thumbing through the chapter until I found the right formula. Then I punched the numbers into a hand calculator and wrote down what the machine read out. No, no, you don't understand. I did not try to figure out each problem first from first principles. The book had already done that. I just looked for the right formula. No, I didn't do an order-of-magnitude calculation next. I just put the numbers into the calculator. Did I let π^2 equal 10? No, I pressed π and then touched the key for squaring. What do you suppose I did in order to multiply by 10? Yeah, I punched 10 and then pressed the multiply button.

There you have it. Straight out and no beating around the bush. Not a pretty picture, you're probably thinking. But if you're expecting remorse, contrition, pleas for forgiveness, forget it. The fact is, I sort of enjoyed it. Furthermore, virtue usually takes time, and I was in a hurry.

Much to my surprise, I relearned a lot of physics using this student method of doing homework. It turns out that to choose the right formula you have to read quite a lot of text around it. One way of reading a chapter is to do a sequence of problems that cover the material in the chapter. Each time you turn back into the text to find a formula, you are asking the text to answer a specific question for you. In security parlance, you have a need to know. No author could hope for a more purposeful reader. No author could invent a better way of ensuring that the chapter gets read, at least in part. Assign enough homework problems and you have an interactive text.

When it comes to doing numerical solutions at the blackboard, I like to be theatrical about zeroing in on answers. In class I scorn the use of a calculator and get order-of-magnitude results with outrageous approximations. You can carry that too far, though. Last month I asked a large lecture class for help in carrying out a calculation that I was about to do. On the board I would do my usual bravado stunts with powers of ten. Since for our particular purposes that day we needed three significant figures for a comparison, I asked volunteers to follow the calculations with their calculators and give me the truth at the end. When they read off their numbers, no one agreed with anyone else. The kids couldn't carry out a series of multiplications and divisions without making mistakes. I had gone too far.

It had never occurred to me that these students couldn't use their little electronic marvels. To be sure, slide rules used to require some skill, but all you have to do with these new things is press buttons. And all the buttons are labeled. Apparently, however, some students do need drill in operating their calculators. If we don't provide the exercise, and the students go on to become engineers, our bridges may collapse. Maybe the kids really do need lots of plug-in problems.

I have not completely abandoned my former ways. Yes, in doing all those problems I did just look up the right formulas, and yes, I put the numbers immediately into a calculator. But, remembering my upbringing, I then looked at each number when it came out. I checked its reasonableness, and the units, and sometimes did the calculation over again in a different way. Using the calculator, of course. When it comes to pressing those keys I'm as shameless as the next guy. Once you've fallen from grace, you might as well wallow. As for looking up formulas, if you can't lick 'em, join 'em.

Hey, kid! Wanta learn physics the easy way?

March 1981

t was bound to happen. The brave new world has arrived, right here in our editorial offices. The Association has purchased a computer-word processor that can do the books, keep the lists, and type our letters. We can store paragraphs in its memory and then call on the computer to assemble them, add a proper salutation, and type it all on good bond paper.

There are a lot of letters that go out of this office—queries, responses, invitations to write an article. Needless to say, each one is personally and carefully composed, but some routine phrases are inevitable. Even editorials have their seasonal routines. Maybe I can prepare enough standard paragraphs to fit any occasion. As the deadline for an editorial approaches each month, I need only press the right buttons on the word processor and out will tumble a finished masterpiece. The machine can choose an opening grabber, pick middle paragraphs at random, and finish with one of the stored conclusions. Of course there's always the danger that the editorial paragraphs might get mixed up with our stored responses to manuscripts. Here's a sample of how it might work.

Spring (summer, winter, fall) is here and once again we physics teachers face new challenges.

The time has come for concerted action in order to save technological education in the United States.

Whatever happened to the cooperation that we used to have between physics teachers and research scientists?

Your thorough and lengthy article on "A New Disproof of Newton's Second Law" is perhaps a little too sophisticated for *The Physics Teacher.* Why don't you try the *American Journal of Physics?*

The facts are straightforward and rather frightening. At no point in the nation's history have more than 20% of high school graduates taken a physics course. During the last few decades, as more and more students remain in school through the 12th grade, the fraction of students taking physics is now no more than 1/5. Statistics can only begin to tell the story.

Thank you for letting us read your lucid article entitled "Toward an Understanding of the Possible Correlation between Class Size and Success in a Competency Test of Bloom's Right Hand Rule." Unfortunately none of our type fonts have sufficient X^2 symbols. Why don't you try the *Journal of Research in Science Teaching?*

What are we to do when only one third of our students can find z, given that $x = y/z$? Are we to conclude that there is a threshold intelligence required for physics and that two thirds of the population fall below it? Or must we point a questioning, if not accusing, finger at math instruction in the lower grades?

Your clever article on "Kittens in the Klassroom" is certainly provocative, but perhaps too technical for our audience. Why don't you try *The Science Teacher?*

As inflation drives up the prices of texts and laboratory equipment, it also forces restrictions on school budgets. In times of increasing costs and decreasing enrollments, we must be prepared to make do, go it alone, and practice the ingenuity for which we physics teachers are justly famous.

Thank you for sending us your brief note on "The Upwelling of Down Drafts in the East Indian Archipelago." While it is true that your discussion of the Coriolis effect on pp. 30–34 would be of interest to our readers, we fear that we cannot run the accompanying expedition photographs in a journal that frequently ends up in subscribers' homes. Why don't you try the *Journal of College Science Teaching?*

Finally, then, our course of action is clear. If physics is to remain alive and well in our schools, it's up to us.

In conclusion, let us remember the somber warning of John Dewey, "Beware the Latin teacher, bearing obsolescence."

One final word. We are only human but a word processor is a machine.

May 1980

Lo, the Vanishing Physics Teacher

T he last week before school opened I received three desperate calls from local high schools, all looking for physics teachers for this fall. These were good jobs being offered. Straight physics teaching in suburban schools. In New York City, they have been short of trained physics teachers for several years. There has been a steady loss of such teachers. Meanwhile, the number of high-school students wanting to study physics, at least in New York State, seems to be going up.

How many high-school physics teachers are there in the United States? The numbers are hard to pin down. The National Science Teachers Association lists about 18,000 who have indicated that they have taught physics or can teach physics. It appears from test mailings, however, that most of these are not primarily trained in physics and do not really consider themselves "physics teachers." We figure that about 4,000 copies of our magazine each month go to high school teachers or high school libraries. Our highest estimate is 5,000 trained high school physics teachers, teaching physics as their prime assignment. The number could be as low as 3,000. We are an endangered species.

Why aren't we trained more? The jobs are waiting, and the jobs are some of the best in teaching. The physics teacher usually deals with the brightest kids in school, and faces fewer discipline and motivation problems than in any other field. Salaries are no longer disgraceful and in some places are even respectable. Opportunities continue to exist for in-service professional learning and accomplishment, both locally and through national organizations.

Apparently college students don't believe high school physics teaching is a good career. Around here, most undergraduate physics majors think that teaching jobs are unavailable. Indeed, the schools are saturated with history, gym, and elementary school teachers, but not with physics teachers. Available or not, high school physics teaching jobs don't sound attractive to most of our undergraduates. Apparently the students attracted to physics were encouraged to think only of a research career. Practically no one comes to us from high school with the ambition to go back to teach physics there.

There's another reason we physics teachers are having trouble propagating ourselves. In the colleges, we've pretty much given up trying. Throughout the northeast, at least, departments that used to have teacher training programs have lost interest in them. Usually, only one or two people in each department took an interest in teacher preparation or had any familiarity with schools. Because of age, or frustration, or tenure problems, many of these specialists have now turned to university assignments with greater respectability. Students aren't knocking on our doors for teacher training, and we aren't out there beating the bushes.

Sometimes it seems as if there aren't even a few thousand high school physics teachers out there. The majority of manuscripts that we receive come from college teachers, though we will bend over backward to publish something from the high schools. Nevertheless, when we conducted a reader survey by mail last spring, the percentage response of high school teachers was far better than from our college subscribers. Apparently the physics teachers in the schools do care about the profession, and surely the ones in the colleges ought to.

There's work for both groups if we're going to replenish our numbers. The high school physics teacher should be a prime recruiter for future teachers as well as for future researchers. Personalities of these two types are usually quite different. Plant the notion that high school physics teaching is a career to be considered. Help a likely candidate to find the right college with a good training program—if there are any left in your part of the country. The colleges that still have such programs should advertise. Indeed, let me know the details of good programs and we'll publish them in the magazine—free.

As for whether or not physics teaching is a good career, let your students read the articles in our series that started last month—The Real World of Physics Teaching. We'll have descriptions of widely different situations, in both high school and college. Our profession has both delights and special problems. Let's tell it like it is to both ourselves and our students. Maybe more will join the ranks. If they don't, the few of us left may want to learn to teach Latin as a hedge against obsolescence.

October 1979

T hose who live close to the land know the rhythms of the seasons. There is an odor of spring even before the first crocus swells the earth. In August, while the evenings are still hot, the night sounds of locusts warn of approaching fall. Physics teachers, also, are attuned to the seasons. Both our students and our topics change in annual progression.

In most introductory physics courses, vectors begin before summer ends. September is a season for balanced forces and for the description of position as a function of time. Under blue October skies we teach dynamics, while November rains bring conservation of momentum and energy. Fifteen study days till Christmas heralds the advent of thermodynamics. In ringing out classical mechanics, January rings in optics and wave motion and the hint that light is subtle and dual-natured. Winter is the time for electrostatics. Our sweaters cling and sheet lightning can be seen as we draw up more blankets in the night. But spring comes at last, with charges flowing through wires and the first tug of magnetic fields. Atoms stir and solar photons linger into evening. Quarks are heard in the land (though never observed), and before summer comes, all the expanding universe will be summarized in time for final exams.

Each fall brings a new crop of students, just like last year's and yet not quite the same. The quiet boy this year is not brilliant after all. This time the female junior heartbreaker turns out to be a hardworking saint. But they're all younger than they used to be. And not so well prepared, it seems. In fact, nowhere near so well prepared as they were ten years ago, when we were younger. Apparently these students retained nothing from previous science courses and there is no evidence that they ever studied algebra. Nor is there much indication throughout the year that they really understand what we're trying to teach them. Most will memorize enough to pass our standard tests, but very few will be concerned with the subtleties that intrigue us and make the study of physics a philosophical adventure. Yet clearly the seniors are more capable than the juniors, and the juniors are much more mature than the sophomores. It must happen during the summer.

Is there no progress for us, from one year to the next? Or are we, in the Buddhist tradition, bound to an endless cycle of vectors to atoms to vectors, and young strangers to old friends and back to young strangers? Actually, the topics do change, or they can and they ought to. Through summer institutes or our own hobby studies we should review and renew at least one topic every year. If we are bored by vectors, then why not learn about tensors—for ourselves, if not for our students? Why not build one new demonstration device for a topic that has lost its excitement? Perhaps for one topic next year we could try a radically different teaching technique.

Of course, at this season of the year it's hard to get excited about the problems of next fall. It seems sufficient if we can ever get through the next few weeks. But summer will come. One way or another the course will be over, the grades will be handed in, and another group of students will have had physics. By then we will have had it too. Particularly with students and school administrators. For us another cycle will be finished, but our students will be spinning off on other paths that do not repeat. They'll remember us long after we've forgotten them. Their feelings about physics and science will have been permanently altered, one way or another. Sometimes in our annual circling we're visited by students from cycles of the past. We find that there is, after all, a continuity and progression. Our attitudes and styles, if not the facts we taught, have been modified and adopted. We have made a difference.

It only takes a few such hints that our efforts do produce changes now and then to lift us out of the cyclic year-end desperation. In a couple of months the students will look new again, and the course may look more exciting, and once again we'll be ready to be teachers for all seasons.

May 1979

Happy New Year! For a teacher, that means September, of course, not January. Unless we were unlucky enough to be involved with summer school, we have had two months to get over our annual year-end loathing of students. Now we can face this year's crop with patience, good humor, and great expectations. They look younger, somehow, but no doubt we will all age together.

Curricula and texts for the year have been chosen. Our students are slated to go marching through a series of topics, one building on another in a logical sequence known as the physics course. In most classes there will be some vector problems, elaborate attempts to simplify motion enough to demonstrate Newton's laws, the usual excuses for the very special and restricted definition of work, and some attention to thermodynamics, or at least heat. In the second semester there will be waves, electricity, and magnetism, and, if there is time, some study of atoms, molecules, and nuclei which will be called modern physics. Your sequence may be a little different, but this one would surely satisfy a physics requirement.

Suppose it weren't like that. Suppose that there weren't any textbooks, or any course guides, or any outside exams to be met. Forget about the supposed need of some students to learn about certain topics in preparation for further physics courses. What would you do with these kids for the year? How many weeks would you spend on vectors, and how complicated would you make the experiments in Ohm's law? Here's a list of things that I would like to teach students. Your list would no doubt be different, and maybe neither list would look like physics.

First of all, and last of all, and at all the times in between, I would teach a quantitative approach to problems. I would always insist on answers in terms of actual numbers (for this purpose π is not an actual number). Furthermore, units should not only be attached to the number, but the student should be able to describe the size of that many units in terms indicating recognition and experience. (Is 10^8 joules large in this context? Is it expensive?) This fetish for the quantitative does not involve the use of computers. Far from it! Every class topic, scientific or otherwise, would be analyzed in a series of successive approximations.

First, the order of magnitude calculation. (With the mechanical pump there are 10^{16} molecules per c.c. The gross national product is 10^{12}.) Then, if it is needed, proceed to the back-of-the-envelope calculation to get one significant figure. There might even be a rare occasion when there would be justification for using a slide rule and getting three significant figures. Day after day, in formal lecture and in casual conversations the students would learn to clothe their experience with numbers. (How do I love thee? Why, yes, let me count the ways.)

I would want to make sure that students learned their place. It is an ancient quest. (What is man, that Thou art mindful of him?) In the class called physics I would emphasize the physical place of humans, those fantastic assemblages of particles from the microworld, immersed in a cosmos of time and space that in some ways can and should be comprehended. If I discovered that my students didn't have a good first approximation to that comprehension, I'd be tempted to teach that even if I had to leave out the lessons on vectors.

Physics seems to be the only class where students learn about functional dependence of variables. I would enhance that mission with descriptive terms and with graphs. (Why should Newton's gravitation law contain the inverse square? Could the negative exponent really be 2.00001? Why do the tides depend on the inverse cube?)

Perhaps, if it weren't for the curriculum, I would forget calorimetry or lenses, and spend more time on interpreting phenomena in terms of only four interactions (gravity, electromagnetic, strong nuclear, and weak) constrained by only a few conservation laws.

The standard curriculum doesn't contain a chapter on curiosity, and perhaps it can't be taught. It's never mentioned explicitly in the seven steps of the experimental method. In my own experience, though, it's the driving force behind real science. Perhaps the classroom could reflect some of that spirit if the teacher could find some reason at least once a week to admit, "I don't know. How will we find out?" This is just one example from what has unfortunately come to be called the affective domain. Such things are not only hard to teach, they just aren't testable with the standard curriculum.

But, of course, there is a curriculum, and a chosen text, and guidelines imposed one way or the other. Even if there weren't, it's tough to design and provide lessons and experiences de novo. Maybe there is a way, however, for each of us to teach his private course within the public physics course. Draw up a hidden agenda. Actually write out the skills and knowledge you realistically think that you can teach and that the students can learn. Rank them according to your own priorities. Perhaps there's a perfect match between the course you're slated to teach, and your own list of what's important. If that's not the case, why not supplement and shape the curriculum with your hidden agenda. It happens anyway, of course, because the hidden agenda in every classroom is the teacher. Your personality, and ambitions, and interests, and expectations are prime topics of study by every one of your students. If you make your own agenda explicit for yourself, and keep it in mind throughout the year, your students may find a fair fraction of it embedded in the standard curriculum.

September 1974

The Moral Equivalent of Physics

Before I first taught physics in a high school, I talked with my predecessor about his methods and problems. He was eager that I continue his program of stressing the social implications of science, particularly with regard to the dangers of nuclear bombs and power plants. I was polite but not persuaded. It seemed to me that the prime job of the physics teacher was to teach the principles of physics. Applications should be considered only if they served as tools for understanding basic principles. Sociology and government were problems for other classes and for teachers who wanted to avoid the hard discipline of preparation by staging impromptu moral wrestling matches in their classrooms.

I think that my attitude was characteristic of that of most physics teachers in those days—certainly of those in colleges. I remember once receiving a school science book entitled *Teaching Science Through Conservation*. A theoretiker friend of mine picked it up, exclaiming with delight, "At last somebody has done it! The conservation laws *should* serve as the basis for introductory physics." Alas, this book was concerned with saving wetlands, whales, and bristlecone pines.

There have been precedents for scientists turning away from pure research and fashioning their meter sticks into swords. Practically the whole physics establishment cooperated to turn the second world war into the war of the physicists. They turned to applied physics with zest, and in the end turned gladly back, a little shaken by what they had done.

It appears that we are in another crisis now. We really are. The president asserts it; the supermarket prices confirm it. For Americans, the days of cheap energy are over. The search for national energy independence has been declared to be the moral equivalent of war. Maybe, once again, physicists and physics teachers should heed the trumpet call.

The first question is, do we really hear that trumpet call, and is it the same one our students hear? A fair share of the American public still thinks that the fuel crisis was caused by manipulations of oil companies or foreign potentates. It may well be that our temporary problems are caused by the fumbling of management or government, but in the longer run of the next few decades, the exponential function has already trapped us. Our first obligation as interpreters of the language of technology is to convince ourselves and others that there have to be changes in our use of energy. Our increasing troubles are not caused by human conspiracies, but by the relentless scramble of human desires digging into finite natural resources.

Unfortunately, the situation is complicated. The proposed cures are complex and contradictory. Technical solutions have sociological implications. Solar energy can lead to greater centralization of technological control if satellites are used; on the other hand, solar energy can lead to decentralization of control if roof top collectors can be used. Physical efficiencies are frequently not so important as economic efficiencies, and these in turn may depend on the extent to which we exploit commonly held resources or mortgage our children's future. It may seem like science fair fun to build solar collectors, and there's lots of physics to be learned doing it, but we must also learn how to calculate economic realities. There are no simple solutions to our energy problems. That fact may be as important to our curriculum as Newton's second law. Who better than the physics teacher can teach about the complexity of real world problems?

Who better than the physics teacher can teach the vital details that require quantitative calculations of concentrations (is 1 part per million dangerous?) or surface densities (how dispersed is 1 kW/m^2?). What better analogy is there for heat flow from a house than electrical network theory? (Doubling 10 Ω that is in parallel with 1 Ω makes little difference in a circuit; in a house with loose caulking and weatherstripping, doubling the insulation may be futile.) It's in the physics class that students should learn the difference between energy and fuels, and learn that it may take more energy to produce some fuels than they are worth. (Where do you get the energy to produce alcohol or extract shale oil or make hydrogen?) Unless people learn the difference in physics class between the first and second law of thermodynamics, they won't really understand heat pumps or know why oil furnaces, according to the second law, are grossly inefficient.

Finally, for 30 years people have assumed that physics teachers are the local experts on nuclear energy. Most of us aren't; maybe we ought to be.

Marc Ross from the University of Michigan has proposed that we establish programs (perhaps at the master's level) to train people to be "house doctors." Assuming that they would make house calls, they could use modern instrumentation along with the folklore of experienced builders to analyze existing heating systems and insulation characteristics. The subject is not simple, and not all of the instruments or curriculum are at hand. But according to Ross there is an enormous and immediate saving to be made in this type of energy conservation.

Perhaps individually, and in our organizations, we should take up the new hobby of learning and teaching about the complexities of energy conversion. In this time of war, or moral equivalent thereof, we physics teachers must learn to measure the R values of the insulation in our own ivory towers.

September 1979

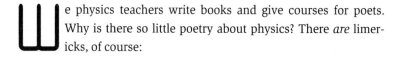

We physics teachers write books and give courses for poets. Why is there so little poetry about physics? There *are* limericks, of course:

> *There once was a flyer named Wright,*
> *Whose speed was much faster than light.*
> *He set out one day*
> *In a relative way*
> *And came home the previous night.*

Here's another one that contains a brief reference to a standard physics demonstration:

> *A leacherous tenor named Squire*
> *Breathes helium to make his voice higher.*
> *But in spite of his fame*
> *The sopranos all claim*
> *He's unspeakably base in the choir.*

Unfortunately, most of the good physics limericks are inappropriate for a family magazine.

A few poets have dealt seriously with physics. In the October 1973 issue of *The Physics Teacher,* Sister Martha Ryder gave us an example. The late 19th century English mystic, Gerard Manley Hopkins, linked religious poetry with precise and perceptive descriptions of natural phenomena. Sister Ryder chose a selection from "The Blessed Virgin Compared With The Air We Breathe."

> *Again, look overhead*
> *How air is azuréd . . .*
> *Whereas did air not make*
> *This bath of blue and slake*
> *His fire, the sun would shake,*
> *A blear and blinding ball*

> With blackness bound, and all
> The thick stars round him roll
> Flashing like flecks of coal,
> Quartz-fret, or sparks of salt,
> In grimy vasty vault.

In a sensitive gloss, explaining how she uses this poem in a physics class, Sister Ryder notes, "By this time I am almost in tears because I like the poem so much . . . ."

Wouldn't it be great if we could link heart and head with all the physics topics that we teach? What other poetry is available? There are a few rare examples from other eras. Two thousand years ago, the Roman poet Lucretius, wrote a monumental treatise in hexameter verse on the atomic theory of Epicurus. Here are a few lines translated from *De Rerum Natura—On The Nature of Things:*

> . . . . Bodies, again,
> Are partly primal germs of things, and partly
> Unions deriving from the primal germs.
> And those which are the primal germs of things
> No power can quench; for in the end they conquer
> By their own solidness; though hard it be
> To think that aught in things has solid frame;
> For lightnings pass, no less than voice and shout,
> Through hedging walls of houses, and the iron
> White-dazzles in the fire, and rocks will burn
> With exhalations fierce and burst asunder.

In the June 1977 *Scientific American*, Margaret Byard wrote about the poetic responses to the Copernican revolution. In the 19th century, poets were fascinated by the new discoveries of science. Here's a selection from George Herbert, a contemporary of John Donne and Galileo:

> Philosophers have measur'd mountains,
> Fathom'd the depths of seas, of states, and kings,
> Walk'd with a staffe to heav'n, and traced fountains. . . .

Sad to say, in most cases when poets deal with physics, or science of any kind, they are cool if not downright antagonistic. There is a long tradition for this attitude. In the 38th chapter of Job, God speaks out of the whirlwind, saying:

> *Where wast thou when I laid the foundations of the earth?*
> *Or who laid the cornerstone thereof,*
> *When the morning stars sang together?*

Job, who had dared to question fate, could only respond:

> *What shall I answer thee?*
> *I will lay mine hand upon my mouth.*

Walt Whitman celebrated the new Americans, yet took a jaundiced view of science:

> *When I heard the learn'd astronomer,*
> *When the proofs, the figures, were ranged in columns*
>   *before me,*
> *When I was shown the charts and diagrams, to add,*
>   *divide, and measure them,*
> *When I sitting heard the astronomer where he lectured with much*
>   *applause in the lecture-room,*
> *How soon unaccountable I became tired and sick,*
> *Till rising and gliding out I wander'd off by myself,*
> *In the mystical moist night-air, and from time to time,*
> *Look'd up in perfect silence at the stars.*

More recently, John Ciardi implies the same mistrust in his flashy and colorful poem, "My Father's Watch." Donald Holcomb and Philip Morrison used the poem's title for the title of their physics text. The poem starts out:

> *One night I dreamed I was locked in my Father's watch*
> *With Ptolemy and twenty-one ruby stars—*

But at the end, when the protagonist has dared to probe the works, and all hell has broken loose:

I saw my Father's face frown through the glass.

The moral seems to be that we shouldn't ask, and we shouldn't analyze, and, particularly, we shouldn't meddle. Evidently, poets don't understand our enterprise or its joy. Perhaps physicists must be our own poets. Surely we have not only a special understanding, but also a special vision. Henri Poincaré spoke for all of us:

> The scientist does not study nature because it is useful; he studies it because he delights in it, and he delights in it because it is beautiful. If nature were not beautiful, it would not be worth knowing, and if nature were not worth knowing, life would not be worth living.

Herewith we extend an invitation to poets who understand physics or physicists who are poets. *The Physics Teacher* welcomes poetry about physics—appreciative of physics preferably, but at any rate, good poetry. The creative triumphs of modern physics deserve modern celebration.

> *There yet remains the challenge to the void,*
> *The gesture, seeking immortality.*
> *We build. Our temples cover earth.*
> *Their spires reach up, their mirrors seek out.*
> *We redesign the animals, and bridge the streams,*
> *And bring the mountains low.*
> *And not content with earth,*
> *We'll rearrange the heavens*
> *And establish ourselves in the uttermost places,*
> *Singing new songs of creation.*

May 1978

Somewhere between martini and Eastern Airlines chicken—which is to say, somewhere between Atlantic City and Durham—we read two articles about scientists and nonscientists. We were high enough at the time to take an Olympian view of this subject and were disturbed. The first article was a report[1] on one of our combined American Association of Physics Teachers (AAPT)—American Physical Society committees—the Committee on Science Education of the General Public. The other article was a survey of "Some Contemporary Humanists' Perceptions of Science," by William Davenport, Professor Emeritus of Literature from Harvey Mudd College. His essay was written for the newsletter[2] published by the Program on Public Conceptions of Science at Harvard University. In both reports the two-culture theme was raised and exemplified. In both cases it was assumed that scientists were the providers of new and disturbing technology and that for continued support by the other culture the scientists must justify the beneficence of their new technology.

According to the newspaper account, the Committee was told that "The antiscience surge stems from a variety of reasons. Since World War II, science has made possible advances in antibiotics, computers, space vehicles, jets, radar, transistors, and lasers. But scientists, too, made possible the atomic bomb, and, later, the H-bomb, defoliation, and chemical warfare."

Professor Davenport, in a mood of reconciliation with the scientists, cites case after case where painting, sculpture, music, theatre, and even poetry, have benefited from the technology of science. However, in most of these examples the word "machine" is used in the description, always accompanied by quotes stressing an underlying fear that the machines will dehumanize our arts.

Nowhere in the report of the Committee discussion or in Davenport's survey is there any hint of intellectual liberation, or philosophical revelations, or the simple joy of studying science. What have we been teach-

[1]Reported in *Newsday* (9 October 1973) by Newsday Science Writer, David Zinman.
[2]Newsletter No. 5, Program on Public Conceptions of Science, 358 Jefferson Physical Laboratory, Harvard University, Cambridge, MA 02138.

ing people all these years? Apparently we ourselves, as well as sympathetic friends, characterize our activities in terms of transistors and bombs, new stage lights and Moog synthesizers.

Of course, technology and science are symbiotic. We have all heard the new-left criticism that science is never pure. Regardless of the type of research, there are always sociological implications. These implications, however, are grander than the threat of new machinery.

Some of our physics courses try to humanize science by taking a detailed look at the history of scientific discoveries. The favorite topic for examination is the Copernicus–Kepler–Galileo–Newton period. That's a good one because it can so easily be interpreted as a literal revolution of our worldview. Project Physics does a superb job of presenting the scientific details and the philosophical issues of that era. Perhaps students view that story as one out of the past, unrelated to the current activities of science. The Physical Science Study Committee (PSSC) course is built on a subtle theme of the nature of reality and our attempts to comprehend phenomena in terms of man-made models. Evidently the theme is too subtle or the concept too advanced for high school students. At any rate, that's the general reputation of the course. Many people ignore the philosophical base and consider the course suitable only for students going on in science.

Certainly most of our physics texts are more concerned with technology than with philosophy. Thumb through them. Look at the pictures showing applications of principles. Examine the problems and tests. Maybe Professor Davenport and the Committee have the right slant on things after all.

What a pity, though, that our main business isn't described or understood. People go into physics rather than engineering because they are more concerned with finding out how things work than they are in making things that work. Learning the first may lead to the second, of course, but that possibility is not the original or driving motive. We are curious about nature. And it's exciting to discover new relationships. What could be more humanistic than that?

It wasn't just three hundred years ago that our worldview changed. It wasn't just that glorious period of the first quarter of this century. The new insights into the nature of the universe and the nature of humans continue to this day. We certainly hope that the students who come from our classes will have gained the technical competence to master their home toasters and will have sufficient appreciation of the economic importance of scientists so that they will vote for large research expenditure. According to those two articles, that seems to be the right relationship between science and society. We also hope, however, that our students will be able to describe the ranges of sizes and times in our universe and can place humans in those spectra. We hope that some of them will remember physics as a course in the humanities that revealed to them an exciting world, different from any that they had known before.

January 1974

A Good and Demanding Life

t was Career Day at the local high school and we were talking to the students who had expressed an interest in physics research or physics teaching. The group was small. At first we described the daily life of a person engaged in pure research at one of the national laboratories. Interest was high, and questions were directed along rather idealistic lines concerning how discoveries were made. When we turned to the subject of physics teaching, however, the students clearly thought that we were trying to sell them a bill of goods. Some students in the school belong to a club for future teachers. They had thought about being grade-school teachers or high-school English or history or math teachers. None of them, apparently, had ever pictured himself in the role of a high-school physics teacher, and no one had ever spoken to them about this possibility. They were tolerant, bemused by our pitch, but asked no questions.

How would you sell the job of being a high-school physics teacher to your students? The traditional image is certainly not very appealing. Most of us studied high school physics under a man who had taught the same subject in the same way year after year. In many schools he had to teach that way because the subject was strictly prescribed by state or college requirements. Like all other teachers, his professional contacts were limited to those between students and teacher—between teenagers and an adult. Among themselves, teachers were mainly concerned with matters of administration. In most schools, there was only one person who taught physics, and there was no one with whom he could have talked about physics or physics teaching even if he had wanted to. During the summer months, the physics teacher, like most others, had to leave his profession and find a temporary job in some shop or local industry. Maybe things were better for your high-school physics teacher, but that's the way they were when we went to school.

Times have changed! Maybe not everywhere yet, but in many places high-school physics teaching is now an honorable and rewarding profession. First of all, salaries for all teachers are beginning to be respectable. Many opportunities exist for appropriate professional work during the summer. There are summer institutes providing small stipends for additional study, and an increasing number of local and national curriculum workshops that need the help of experienced teachers. Even if the physics teacher is isolated within his own school, there are many ways to work with Science Teachers Associations, or to contribute to science teaching journals.

Instead of being forced to teach the same topics in the same sequence year after year, the physics teacher today has an almost bewildering array of course models from which to choose. There are Physical Science Study Committee, Harvard Project Physics, a new engineering physics syllabus, and several varieties of physical science for eleventh and twelfth graders. The existence of these new courses provides a real challenge. The good teacher must become familiar with all of them. He should also be familiar with the science curriculum developments for the lower grades. Many new junior-high and elementary-school science programs are being developed or are now available. The high-school physics teacher should know about these courses because they ought to affect what he teaches in his physics class. Even more important is the obligation to be a leader in the coordination of science courses from kindergarten through college. The high-school physics teacher can occupy a crucially strategic position in these efforts. More contacts to the colleges and curriculum development projects are available to the physics teacher than to any of the other science teachers. On the other hand, being part of a school system he should be a natural contact for scientists wishing to work with the schools. In the continual assessment of the K–12 science program which every good system should undertake, the physics teacher has an obligation to advise and suggest. He should know in detail what the various elementary and junior high courses provide and should be prepared to lead in the in-service training of nonspecialist science teachers.

The profession of high-school physics teaching has changed in many places and can change everywhere. Besides the traditional personal satisfaction with successful teaching, there are widespread opportunities to engage in other aspects of the profession. It can be a rewarding career, but the increasing rewards have brought with them serious obligations. That is the way a profession ought to be.

February 1968

Mysteries

The new national standards for science education are due out any day now. The several earlier drafts emphasized that these are standards for teaching and testing and school organization as well as for science topics. Nevertheless, it's the problems of content that have received most of the attention. One of your favorite topics or one of mine is sure not to be included. In the September 1993 editorial I gave a sample of the sort of topics that I would include. These emphasized quantitative skills, a subject almost completely lacking in the first drafts of the national standards.

In the Letters column of this issue we have a note from Reuben Alley, decrying a syllabus for a college physics course that leaves out electricity and magnetism (E & M). Without E & M, he says, it isn't physics. I can sympathize with that, and would add other topics. For instance, without Newton's laws or the behavior of light or energy exchanges, it isn't physics.

There's another physics topic that I hear about almost every week in our colloquium at this research university. But it's not in the Table of Contents of any physics text I know. The topic is mystery, the unknown. Sometimes the nature of the mystery is so esoteric that only specialists in the subject can understand what all the fuss is about. In most cases, however, the abyss in our understanding is only a step beyond some standard topic in our introductory course.

Here are some examples of things that still send a chill down my back. First, there's the scale of our universe, in space and in time. We casually do experiments with subatomic particles that, to the extent their size is meaningful, are only 10^{-15} m in diameter. We can measure distances smaller than that by a factor of 10^6. We also do experiments, in a different way, with quasars at the edge of the universe, at a distance of 10^{-6} m. We humans are in between—but we know it. We can measure time intervals of 10^{-23} s. and make reasonable theories about what happened in much shorter times at the beginning of our universe. We measure the age of that universe, with some uncertainty, and the age of our own planet with great precision. Our little lives seem insignificant—but we are the ones who measure those times.

Don't you also feel uneasy about the way light behaves in the region of a double slit? Note that I avoid describing light as "traveling" or "passing through." With both slits open we get interference of the "waves." With only one slit open, the "photon" no longer has to avoid the dark bands forbidden by interference. (If the photon had to "pass through" the open slit, how did it know that the other one was closed?)

If you have a conductor *in* a changing magnetic field, Lorentz tells us that there will be forces on the electric charges, and thus a current if the circuit allows. We can get used to that. Now put a loop of wire around a cylindrical electromagnet and arrange the return path for the magnetic field lines so that the loop remains essentially in zero magnetic field. Change the magnetic field strength and there will be a current in the wire. Why not? It's like a transformer. But there is no magnetic field out there and no Lorentz force. We're all so used to that phenomenon that it probably doesn't even seem remarkable. There is a similar phenomenon involving the reality of the magnetic vector potential. It's called the Aharanov–Bohm effect and astonished a lot of people when it was observed only a few years ago.

A century ago Maxwell and others unified our understanding of electric and magnetic phenomena. In recent decades we have learned how to unify the Weak interaction and electromagnetism. Now the theory of quantum chromodynamics has brought the Strong nuclear interaction into the fold (with a few colorful questions and strong mysteries left). But where and what is good old-fashioned gravity? Didn't Einstein settle that question 80 years ago with his theory of general relativity? But the most pervasive interaction of all has not yet been explained in terms of a grand unifying theory. Even at the introductory physics level, it remains mysterious why inertial mass equals gravitational mass.

Is the missing mass in the universe some kind of "dark matter"? Perhaps neutrinos have the missing mass. If so, they can transform among themselves, perhaps explaining the missing neutrinos from our Sun. And perhaps not.

The province of physics is filled with unanswered questions, but they are good questions. They are asked in ways that can lead to answers. The primeval questions remain. Are we alone? Are we the only creatures in the universe who comprehend, or at least live and propagate? Is there some point to our existence? Such questions lead to the border of our competence, but look how our discoveries during the last century have altered the ambience of the questions and have constrained the possible answers.

The topics of mystery and wonder are not listed in any Table of Contents or in the National Science Standards. What a pity, however, if a student should leave our physics classes thinking that the important topics were $F = ma$ in the first semester and $I = V/R$ in the second semester. Mystery and wonder should not be a separate chapter, perhaps to be left out if time is short. The attitude of marvel should permeate our teaching throughout the year. For every topic and for every formula there is something not obvious, something one step beyond that we do not really understand. We teach a subject that glories in that situation and we should make sure our students know it and marvel also.

January 1996

went out to see Hale-Bopp last night. It's a dramatic show, particularly if you know what you're looking at. I was reminded of other celestial events I've seen. There was a glorious full eclipse of the Sun we witnessed off the coast of Africa, and once, while in a holding pattern at 10,000 feet, I watched a setting Venus turn into a brilliant ruby. On a more homely scale, there is never a day but what the heavens display wonders of fluid motion and optics and thermodynamics. Physics?

Of course it's physics. We are surrounded by it. Our classrooms and laboratories provide the keys to understanding our world. The resonance experiment with a tube closed at one end models the first approximation to the tides in Long Island Sound. The scattering of light in a foggy tank of water shows us why the sky is blue. Our centripetal force apparatus and Newton's law of gravitation combine to explain the orbital motions of planets. The tabletop Van de Graaff mimics the thunderheads. Ripple tanks and Slinkies display waves that are the key to understanding the tremors of the globe we live on.

Do we really want to know why the sky is blue and sunsets are red? Wouldn't it be better just to enjoy the colors, or write songs about them? Deep enjoyment does not come so simply. The songs of great composers or the whales are meaningless without a tutored ear. The sunsets and cathedrals are diminished when we are blind to intricate details. Humans are driven to understand.

What is the natural reaction when you tell people that you teach physics? They shudder. I remember a doctor telling me that physics was his worst subject, but fortunately, he had no need of it in his practice. Physics for him meant experiments with carts on inclined planes and hooking up wires according to a diagram. Usually things didn't work. Why is there this disparity between comprehending the marvels of our universe and teaching grubby details about mechanics and nineteenth-century electricity? The details do have to be learned before you can take the next steps. The introductory textbooks do have to devote most of their space to the fundamentals. Guess who's left to provide the sense of wonder and excitement and awe?

It is an awful universe, in the original meaning of the word. It is now one hundred years since the discovery of the electron. During this century we have learned an enormous amount about our universe. In particular, we have learned that there is an awful lot that we do not know. These are mostly things we did not know that we did not know, one hundred years ago. The scale of our universe has expanded into a micro world dominated by quantum mechanics and to a cosmos of violent phenomena. It is curious, but satisfying, that the discoveries at one end of the scale must be linked with observations at the other end. Particle physics must explain, and be explained by, the scenario at the beginning of time.

So what is so awful about an object in free fall, or a cart rolling down an inclined plane? Well, why does mass cancel out? Why is inertial mass equal to gravitational mass? That's only one small profound step beyond the introductory lesson. Let your students know the implications. A few will be intrigued. Where does energy go when you heat solids or liquids, and why is it different for different materials? Simple quantum mechanics can be invoked even at the introductory level. The two-slit interference effect is filled with marvels at several levels. If your students are not excited by the explanation of adding wave crests from two different sources, show them what happens when the apparatus contains only one photon at a time. As for simple circuits, it's not so obvious why $I = V/R$, and sometimes it doesn't. Electromagnetic induction is not obvious at all. How do you explain induced current in a wire surrounding a changing magnetic field, when the wire is never *in* the field? That's mysterious.

Late one afternoon outside a grocery store, I looked to the west and saw the Sun break through the passing storm clouds. On either side were bright sun dogs. I pointed them out to passersby. "Look, look at the sun dogs up there!" Naturally, they shrank away from me and hurried into the store. Nevertheless, I think that physics teachers have a mission, along with our other duties of taking attendance and preparing students for exams. We know that we live in an astonishing universe, one to be studied with excitement and awe. That's what physics is about. It's up to us to keep calling to our students. "Oh, look, look!" Even in introductory physics, the things to be seen are awesome.

May 1997

Extrapolations and Prognostications

I t's that time of millennium again when we look to the past and prophesy about the future. My own family memories don't go much past the twentieth century, although my wife is a direct descendant of a woman convicted of witchcraft in Salem. Not recently, of course. A century ago my mother was going to school in a one-room schoolhouse with student desks in rows and a blackboard up front for the teacher. In high school she studied physics but there were no student laboratory exercises and only a few demonstrations by the teacher. She didn't like it.

What do you suppose physics teaching will be like in 3000, or even in 2100? Surely there will still be emphasis on seventeenth century mechanics. Surely the Piagetian levels will still limit concept development. Most 18 year olds will still think that the acceleration of a ball at the top of its trajectory is zero. But will there still be those student desks in rows and the teacher up front with the blackboard, or perhaps the whiteboard? As a first approximation, can we predict the future by extrapolating from the past?

Certainly there has not been much change in educational methods during most of this past century. The geometry of the classrooms is much the same as in yesteryear. The textbooks have four-color printing and are much heavier, but the basic style and method of use hasn't changed. The most common form of grading is still a set of questions to be answered during a test period.

Can we therefore extrapolate and claim that our present and past educational style will continue for another hundred years? Maybe not. Computer technology has affected every phase of life including education. Within the past ten years we have seen a revolution in equipment for education labs. There are now electronic sensors for almost all variables that we want to measure in introductory physics, and these can feed into simple computers that massage the data and graph it. The computer Web has opened up the classroom so that students and teacher need never meet in person. The number of people involved in distance learning at college level is growing rapidly. Even in courses dependent on lectures, the teacher can maintain an electronic message board, posting assignments, problems, lecture notes, and even exams.

These recent changes should warn us about the dangers of extrapolating very far into the future. However, we should also remember what happened to previous technological marvels. Seventy-five years ago, when radio was spreading across the nation, there were attempts to make it a medium of education. Everyone was going to hear great lectures from great lecturers. Fifty years ago television was to be the technological fix. As a harbinger of panaceas to come, Sunrise Seminar showed good physics demonstrations performed by Harvey White, a superb teacher. But the revolution didn't spread. In the schools the students still sat in rows and the teacher still stood in front of the chalkboard and talked.

Maybe things will be different this time. The computer and the Web-access systems may make an irreversible change in our teaching and learning methods. The popular view is that students will now have the world's facts at their command. As one advertisement for a search program promises, "Provides information faster than thought." I hope not. After all, we have been able to obtain facts from libraries since the days of Alexandria, and from encyclopedias in English since 1775. The important goal of education is not knowledge of the facts, but the ability to connect the facts. I don't know of any computer program that does this, or teaches how to do it.

For a prediction made in 1900 about the nature of schooling in 2000, extrapolation would have been valid. But progress in computer systems has been so rapid in the last decade that now it seems foolish to predict what schools will be like in another century. However, since neither I nor any of my faithful readers will be around then, I will be bold enough to say sooth without fear of contradiction. In 2100, the learning processes of people will not have changed one iota from their present state. Children will still learn a major share of their knowledge by rote memorization and of their skills by repetitive practice. Physics concepts in terms of grand generalizations will still be learned by no more than a fourth of teenagers, and then only with much cyclic experience. Possibly books will be replaced by coded disks, but there will still be school buildings in which students can be instructed and monitored. There will still be tests to determine advancement. And there will still be a need for teachers to have human contact with their students.

January 2000

Part II CHEER & GLOOM

The students aren't as good as they used to be, and they never were. By the rivers of Academia, let us sit down and weep. On the other hand, our physics students are in the upper quintile of high school students. Many of them seem to like the course and some of them even understand the physics.

Here are a few gloomy thoughts and a few cheerful ones.

Scientific truth should be presented in different forms, and should be regarded as equally scientific whether it appears in the robust form and the vivid colouring of a physical illustration, or in the tenuity and paleness of a symbolic expression.
—James Clerk Maxwell, in a tribute to Faraday

Error is not the difference between your result and the "accepted value."
(Who accepted it? What was their error?)

Whereunto then shall I liken the men of this generation? They are like unto children sitting in the marketplace and calling one to another, and saying: We have piped unto you, and ye have not danced; we have mourned to you, and ye have not wept.
—Luke 7:31, 32

O k, Coach. Here's the new team of hopefuls, trying out for the physics varsity. You've got to let them know what they're in for, instill some discipline, and develop a fighting spirit. Of course you'll do this by making sure the sport looks easy and attractive. They'll have a lot of fun here. Tell them how they can team up with their pals to work together. Eases the work for everyone, and they can spend the time talking over the plays or things to do on weekends. Not much homework practice involved. No reason to scare off the team members. Everyone can play and succeed under the new rules. We'll temper the academic winds to the shorn lambs, poor kids.

Who are we kidding? Can you imagine a football coach proposing such a program? Physics is a tough sport. It requires at least as much dedication as soccer or basketball. It requires at least as much practice as the five-finger exercises for piano. There's no golden road to education. There never was and there never will be.

Remember the old story of the professor who opened the school year by saying, "Ladies and gentlemen, I want each of you to shake hands with the person on your left. Now shake hands with the person on your right. Only one of you will finish the course." There may be something to that.

Perhaps there's no need to be so draconian. But fear is an honest goad for student progress. Of course, you should entice your students with the mysteries of physics and invite them to enjoy the beauties of our world view. But make them work to get there. Nothing so clarifies the academic mind and makes students hungry for learning as the imminence of an exam.

This may be an unpopular doctrine. I may feel a lot better tomorrow if the Sun comes out or if a student asks a good question. Today, however, I am reminded that we have lots of meetings and projects designed to make physics more palatable and to make us more knowledgeable about our students and their problems. Maybe the emphasis should be shifted. Maybe our students should worry about understanding us. If we have good physics to teach and know how to present it, let the kids get down to work. In college the traditional formula calls for two hours of work outside of class for every hour spent in class. There should be a different but similar formula for high-school students.

It's our responsibility as teachers to choose appropriate explanations, to make the examples interesting and relevant, and to design lessons at the right level for efficient learning. It's the student's responsibility to spend efficient time and effort on the assignments and to ask good questions. When I play the pipe, I expect the students to dance. And when I mourn, they'd better weep.

January 1999

was in a holding pattern in our local hospital when a radiologist friend walked by and asked after my health. When I told him that I was scheduled for a thallium stress test, he said, "Oh, we don't use thallium any more; it's technetium now." Five minutes later as I was being wired up, I said to the attending physician, "I understand that it's technetium I'll be getting." "Oh, no," she said. "It's thallium. We hardly ever use technetium."

There's probably some simple explanation, and besides it probably didn't make any difference in the results. Gamma rays are gamma rays. It did occur to me, however, as I was being slid into a large ring machine, that I didn't know how the machine worked. Evidently those gamma rays were going to produce an image of my heart. How? I'm supposed to know about gamma rays, but I don't know how their effects are scanned to form an image. I asked the nurse technician, who cheerily assured me that "it's like a camera." But note! She knew how to make the machine work.

I don't know how most things work. Our oil burner stopped today and I watched the serviceman take things apart, including things I hadn't even noticed before. He had never taken physics in school, but he knew where the filters were and what the flame should look like. In his column and book, *How Things Work,* Dick Crane has explained the innards of a lot of devices. Down in Virginia Louis Bloomfield has made a physics course out of explaining the operation of simple things. He described the course in the October 1997 issue of *The Physics Teacher.*

In most introductory physics texts there is an explanation of electric motors and a simple diagram showing the principle. Have you ever compared the diagram with a real motor? The ones in electric clocks don't look at all like the text version. The ones in kitchen devices or pump motors are much more sophisticated than the ones in the text diagrams. High-power electric generators are turbines that are a far cry from the pistons and cylinders of the textbook Carnot cycle. Indeed, their efficiency is not given by the standard Carnot formula. The apparatus used by Geiger and Marsden to measure the size of the nucleus was very different from the schematics in our texts that describe Rutherford's experiment.

If studying physics and teaching it all these years doesn't equip me to understand the technical world we live in, what's the use of it? I hope—we all believe—that the physics of the introductory course provides a foundation for understanding a wide range of technical applications. We may not know the detailed operation of the devices we use every day—the watch, the carburetor, the computer, the TV—but we have the tools to ask the right questions and to understand the first approximation.

As teachers, there are other mysteries we face. We can't be aware of the day-by-day psychological state of each of our students. We may not even notice if they're asleep, or if they're there at all. Even in one-on-one conversations with students, we have a hard time knowing what they are thinking while we are talking. Of course, we should try to get the students to talk while we listen, but even then it's hard to know what they are thinking, or what model is in their minds.

For high-school teachers, certification requires courses in adolescent psychology, learning theory, and methods of teaching. This preparation is analogous to studying the introductory course of physics. The foundation is supposed to make it easier to understand the applications. College teachers rarely take these courses. Their ability to understand the learning process and to teach descends upon them like grace as the doctoral hood falls on their shoulders. Whether in high school or college, the teacher is expected to be a dual specialist in the subject matter and in the application of the teaching process.

We all know that it is seldom that way. It is usually a straightforward process to explain technology in terms of basic physics. However, it is much more complicated to interpret the efficacy of teaching methods in terms of basic theories of learning. The many factors involved usually prohibit simple conclusions. Good teachers are like good specialists; they may not know why the system works, but they do know how to make it work.

March 1999

Is This Generation Different?

Whereunto shall I liken this generation? They are like children sitting in the marketplace, and calling one to another, and saying. We have piped unto you, and ye have not danced; we have mourned to you, and ye have not wept.

—Luke 7:31, 32

We take as our text this month those chiding words from the New Testament. It's hard to get a response out of this present generation of students. Local high-school teachers have been telling me for some time that in their schools, things are different these days. The kids are nice enough, they are not outrageous in clothes or manners, but they just don't do any work. They're passive. They seem to absorb, but they can't produce.

I certainly see the effect here at the college level, or think I do. A couple of years ago I took my turn at teaching our introductory course with calculus. The last time I did that was in 1965. What a difference! My records show that those earlier students covered more topics and got considerably higher grades than our present ones. Indeed, we have had so many failures and drop outs recently in Physics 101 that we have instituted a one semester "prelude" course to prepare students. In most cases the students in the prelude course have "had" high school math and science. They just don't know how to use it. At this late stage of their lives, we try to teach them how to measure, and graph, and get numbers out of simple calculations.

Perhaps this is just a Stony Brook phenomenon, but I don't think so. In connection with another project, I wrote a description of the situation that led us to start the "prelude" course. This description was circulated to fourteen people around the country who teach introductory physics courses at large universities. Their responses were nearly unanimous—the same conditions seem to exist everywhere. Students are less capable than they used to be in using the standard skills of measurement and calculation.

There's a well known danger in judging today's students in terms of our memories. Today's students never seem as good as yesterday's. It has always been that way. It is reported that Chiron complained that

Hippocrates was not so well prepared as Asclepius, his first premed student. Still, we have the murky evidence of the falling SAT (Scholastic Aptitude Test) scores. There is also a pessimistic report of a group commissioned by the National Science Foundation to assess the science teaching in the schools. They cite a diminishing regard for authority among students, and a general unwillingness to do assignments or to believe that they are worth doing. This malaise among the students is superimposed on a general decline in emphasis on science in the curriculum. Science as inquiry and problem solving is giving way to science as memorization, which is cheaper.

The major problem that we see at every level is that our students have learned words but have no power over them. They know that energy is conserved but don't know how to calculate the efficiency of simple machines. They know that there are atoms, and even quarks, but can't do the arithmetic to figure out the masses or interaction distances. They know that light is a wave (or is it a particle?) but can't measure the wavelength. What's more, unless they're premeds and need the grade, most students aren't particularly concerned. They are only a little annoyed and puzzled that more is required, when, after all, they really "understand" the material.

Even if we were sure that something is different about today's students, we, as individuals, are powerless to do much about it. In many schools these days, teachers are caught between demonstrating to the administration that they're fulfilling their contract and reassuring the union that they aren't overfilling it. Our Association and this magazine provide endless suggestions for piping to your students, but if the students won't dance, should teachers keep playing?

Let us at least mourn to each other. Are your students really different these days? Do you have hard evidence that things have changed, or are we all just getting older? Please let me know. I'll collect the evidence and report back to everyone. If your answers pipe a merry tune, we'll learn to dance again. Otherwise, we can all weep together.

December 1978

No One Kissed the Physics Teacher

I f one of your own children is getting an award, then of course you have to attend the Awards Night. So I marched off last June to witness the annual ritual in the school auditorium. As I feared, the place was packed. Apparently every student in town was slated to be rewarded for something. There were prizes in athletics and English and math and Latin and music and every other subject including physics. As the names were called and the students came forward, there developed a regular orgy of congratulations and hand shaking and general good feelings between students and teachers. By the time we got to the music awards the auditorium was awash with tears and hugs and wild hyperbole. But nobody kissed the physics teacher.

Ask anybody, they'll tell you that physics teachers deal with a cold and logical subject. The abstractions of music can set your toes to tapping, but the abstractions of physics are tuneless. The good fellowship of a choir can bathe students and teachers in harmony, but the competition produced by a short answer quiz in physics leads only to discord. A smartly stepping band brings out the Music Boosters. A snappy physics demonstration may bring out only complaints from the custodian.

Actually, we all know that it needn't be this way. Physics can be exciting, and fun, and even thrilling, if not romantic. It's just that we have a bad press. To overcome the legend we must use salesmanship, showmanship, and lots of public relations techniques. There's a great example of such methods in this month's article, "Physics Olympics," by David Riban from Indiana University of Pennsylvania.

Most science fairs are arranged as islands of individual competition. The projects are usually done in isolation and then are presented on their solitary tables. It's hard to develop an esprit de corps with each entry in grim competition with everyone else. Where would the cheerleaders perform?

The Physics Olympiad creates an entirely different atmosphere for a science contest. Team effort and group projects become important, with school honor at stake. Whether or not anyone learns any physics, at least the reputation and atmosphere of the physics class will change.

(As for the learning benefits, they are probably equivalent to the physical health benefits of varsity sports, but less painful.)

It's the group nature of activities that makes students feel comfortable and wanted. That's why most alumni remember their extracurricular projects more fondly than their scheduled classes. You don't have to mount an Olympiad to bring in the students, nor do the group projects have to be completely separated from classwork. A whole class might prepare an exhibit of physics phenomena for a younger group or a hall demonstration. A whole class might build a telescope, or try Galileo for heresy, or take over some of the technical aspects of the school's photography or public address systems. If you have developed such group projects that are instructive and fun for both you and your students, send a Note to *The Physics Teacher* and let the rest of us know.

We physics teachers realize that the study of physics contains its own intrinsic rewards. What could be more profoundly exciting than the study of the microstructure of matter, or the mysteries of cosmology? Back in the introductory classroom, however, the study of vectors and Newton's laws might benefit from a little razzmatazz. You may not want your students to kiss you goodbye at the end of the year, but there's no law of physics that says that every eye must be dry on that last day of class.

November 1976

Wealth and the Physics Teacher

urely, the title is an oxymoron, a contradiction in terms. Who would start out to make their fortune by becoming a physics teacher? Ah, you say, I know the purpose of this gambit. The editorial is going to turn into a paean to the joys of learning and teaching, even though one is mired in ennobling poverty.

Nonsense! I'm talking cold, hard cash. I'm talking about living it up, on the footstool if not the lap of luxury. I'm talking about advice to questing youth with their eye on the bottom line.

I know the folklore. Teachers are overworked and underpaid. Any of us could leave our ungrateful institutions and go to work in industry at twice the salary. The myth is most poignant when applied to schoolteachers, as opposed to university professors, who, as everyone knows, earn two to three times as much, though still a pittance.

We got a letter a few weeks ago from a high-school teacher who is resigning from the American Association of Physics Teachers because we raised our membership dues slightly. She says she earns one-third to one-quarter of a university professor's salary, and that we should have a more modest category of membership for high-school teachers.

I felt real bad about that letter. It came as a shock because it doesn't at all reflect our local situation. Quite a few of my former students are teaching high-school physics here on Long Island or in the New York metropolitan area. Those who have been out for twenty years or so are earning as much or more than I am. We're all doing all right compared with some of our colleagues or students who went into industry and are now looking for jobs. A typical starting salary on Long Island for a high-school teacher with a bachelor's and no experience is $32,000.* With a master's and 10 years' experience, the salary is around $57,000. All salaries are determined by a step matrix that gives credit for longevity year by year, and for graduate studies. Top salaries, without administrative responsibility, are above $80,000. The average salary for a public school teacher on Long Island is $63,000. Of course, you should see our

*Note that this editorial was written in 1995. Salaries have increased dramatically since then.

living expenses! Real estate taxes are ferocious. They have to be, to pay those salaries.

Are we in a particular Eden here, so close to The City? I made phone calls to friends far and near to inquire about their financial well-being. In most regions around the country, the *relative* status of physics teachers in high school and college is about the same. In rural regions both are earning less than in big cities, but in most places the high-school teacher with master's and 10 years earns about the same as a fresh Ph.D. in the local college, with about the same prospects for future increases. In some southern states, particularly in Texas, things appear to be different. In exchange for the tornadoes and warmer weather, our southern brothers and sisters (indeed, a higher ratio of sisters) who teach in high school are earning considerably less than their age cohorts who teach in college. The salaries in colleges are slightly less than those up north, although it appears that nationwide competition serves to normalize college salaries.

So here's my advice for youth just starting out. If you want to make your fortune, get your teacher's certificate along with your bachelor's degree. Start teaching immediately in a pleasant suburban high school in the northeast. Take additional summer and inservice courses to get a master's degree within five years. Then, at your leisure, take an extra course each year in order to continue your diagonal route across the salary matrix. Eight years later, notice that one of your classmates who went on to graduate school has just become an assistant professor at your local college at a salary that is about the same as yours. Pity that she had to scrape along on an assistantship for those eight years, and is now at least $100,000 behind you.

I still feel bad that we have lost a member because of our dues, but I think the remedy is to increase salaries everywhere (except locally where I have to pay those high taxes). To do so, it might help if the nationwide situation were better known and publicized. I would like to hear from readers everywhere about their local salary scales, particularly the comparison of high-school and college salaries. Let me know, and I'll report back.

I am reminded of the biennial investigation of the state university by the legislature out in the midwest. They got this here professor on the stand, and the legislator asked, "Now then, professor, what sort of workload do you have?" The professor answered, "This semester it's nine hours." "Well," said the legislator, "that is a long day, but then the pay is good, and it's easy work."

January 1995

Saying the Sooth

R ecently I was asked to give a talk on the nature of physics teaching in the twenty-first century. At first I was abashed. Who can predict a hundred years into our busy future? But then I thought, who can do it better than I? Surely, there will be no one in the audience who one hundred years later will gainsay me. Thus emboldened, I ventured to say the sooth.

I took for my guiding principle Ecclesiastes 1:9—"What has been is what will be, and what has been done is what will be done; and there is nothing new under the sun." Let's see how valid that principle is if we apply it to the past. Recall the nature of physics teaching in 1897. What if I had been asked then to predict the nature of our craft one century later. I would have looked around and said, "As a first approximation nothing will change. As there is now, there will be then a rectangular classroom, with students seated in rows. On the walls will be chalkboards, which we have had since Joseph Henry first introduced them sixty years ago. Standing at the front of the room, will be a teacher, talking. Because of the recent efforts of Edwin Hall we now have and will have then physics apparatus for student experimentation on Newton's eternal laws, on electric circuits, hydrostatics, and, of course, on simple machines." You see how successful I would have been at predicting the future! I missed only with regard to hydrostatics and simple machines. We do not teach twentieth-century youth about pulleys and siphons.

Fallacy! What if I had applied this principle to predicting the future of physics, as opposed to physics education? In 1897 the electron was being discovered. Ahead of us lay a century of relativity, quantum mechanics, nuclear structure, the Big Bang, transistors, and lasers. It's a different world because of those discoveries in physics. But what about physics teaching and theories of learning? Ahead of us lay the demise of the faculty theory of learning, the rise and fall of Progressivism, Fletcherism, Essentialism, Piagetianism, and Constructivism. Has nothing fundamentally changed because of those revelations about teaching and learning? No. Not yet.

Actually, there is very little new in these theories. I appeal to my guiding principle. Piaget, while he lasted, did a great service to education by pointing out and buttressing what was already known about the nature of children and their conceptual development. However, one hundred years before Piaget, Joseph Henry described the general stages of youthful understanding and the implications for instruction. Two hundred years before that, John Comenius taught that ". . . like Nature, education must take care to get the right timing and the right sequence."

The current brand new thing, though fading, is Constructivism. (*Knowledge in pieces* is taking over.) Learners must construct their own understanding, if not their own reality. It's just like Johann Pestalozzi said in 1810, "A person gets knowledge by his own investigation, not by endless talk about the results of art and science. A person is much more truly educated through that which he does than through that which he learns secondhand." He insisted that education should be child-centered, with allowance for individual differences.

If not new theories, then surely our new technology will change the methods of teaching. We now have MBL (microcomputer-based labs), and by the year 2000, by presidential fiat, all classrooms in the nation will be able to surf the net. The whole world of facts and porn will be at the fingertips of our students. Technology may indeed make a difference, though not by surfing the net. The new instrumentation is important because it changes the time base during which we can make measurements. It is hard to measure $\theta(t)$ for a pendulum with a stopwatch. The action is too fast. It's easy with a photodetector and electronic timer. The black boxes of our instruments will even analyze the data for us, but that's not necessarily good. For advanced students the computer provides the tool for tackling nontrivial problems in all branches of physics—but not at the introductory level for most students.

It is also possible that the computer can provide the bookkeeping we need to provide more individualized instruction for all ages. There have been many attempts in this past century to create systems of individualized instruction. All have succeeded, but all have died off primarily because of the complexity of measuring and recording and reacting to each student's progress. The computer can do these things, even in its present form.

The major problem of education in any age is how to get the student to work harder while requiring less work from the teacher. I will now boldly predict, with little fear of contradiction, that the continuing revolution in technology will change the nature of our schools. They will become more like libraries or museums with workrooms. The teachers will be more like tutors or reference librarians. The students will work on projects, sometimes together, sometimes alone, facing a progression of tests or demonstrations of competence. I also predict that there will be a never-ending series of new learning theories, each fad having a peculiar name and lasting about ten years but exerting very little practical influence.

In 1871 President Garfield gave a talk at Williams College. In paying tribute to Mark Hopkins, the great teacher and long-time president of the college, Garfield said, "Give me a log hut, with only a simple bench, Mark Hopkins on one end and I on the other, and you may have all the buildings, apparatus and libraries without him." In 2097 the buildings may be different, the computers smaller and faster, the adherents to the latest cult in learning more devout, but we will still have need for that simple bench and a concerned and wise teacher on the other end.

April 1997

Part III   FATHERLY ADVICE

I t may seem presumptuous to offer fatherly advice to experienced teachers. Still, of my four score years and ten, eighty will not come again. So, I am allowed.

Some of the advice in these editorials is highly personal and may not apply to your situation. If you teach summer school, you may not welcome suggestions about staying fresh in the fall. If you have no space for labs and must cart demonstration apparatus between classes, you will not appreciate a sermon about hands-on physics.

However, there is an absolute requirement that applies to all of us. You must like the students you teach. You must enjoy their humor, their vitality, and even tolerate their music. For good or for ill, you are a role model—a friend but not a pal. You must not only like students; you must also like physics, and the students must see that you do. For many of them, this will be their first contact with a scientist. Open the windows for them into that other world. Your mission is not to teach them everything you know but to prepare them to go onward without you.

If a thing is worth doing, it is worth doing well,
enough for the purpose at hand
(and it is surely silly and probably wrong to do it any better than that)

Precision is expensive. Don't get it if you don't need it.

T
he rest of the world has calendar years and fiscal years, but we have school years. Summer has gone. Happy New Year!

The theory is that we have had two months of lying fallow, replenishing our energy, regaining our equilibrium, reviving our sense of humor. Now we can face again the eager students, hungering for knowledge about their physical universe. We can strew before them pearls of wisdom, artfully constructing the interactive lessons so that they will want to gather the gems and make them their own. 'Tis a consummation devoutly to be wished.

In the meantime, new years call for resolutions. Even though we break each resolution immediately, it is still the time-honored thing to list them. The exercise itself is therapeutic. To see where we are going, we must consider where we are and where we have been. Each of us must draw up our own resolutions, but here are some items from my list.

1) *Learn more physics.* For myself, there are two goals here. One is to become more familiar with and/or to keep up with recent developments in physics. The new physics is coming thick and fast. Neutrinos apparently can change into each other and so necessarily have mass. (Why necessarily?) The universe went through an inflationary expansion at the beginning. (For how long and to what radius?) The other goal is to learn or understand better some introductory physics. For flowing liquids we have both Bernoulli and Poiseuille. But which one explains which features of the flow of water in a hose? Which one explains an aspirator attached to the nozzle of the hose? In the old days the operation of radio tubes was apparent. You just had to look into the glass envelope and see where the electrons were coming from and going to. I don't have that same understanding about transistors. I keep getting them confused with those little arrows in the wiring diagrams. I resolve to learn enough about transistors so that I could make one.

2) *Learn more about technology.* After all these years I still don't know how ball bearings are made. Then there is that magic powder that consists of 1.00-μ spheres used for electron microscope calibration and for the Millikan "oil drop" experiment. How do they do that? Even if I

knew how transistors work, I still wouldn't know how to fabricate thousands of them on a silicon chip and get them all connected. I press a key of this computer and the letter b appears. How many flips have flopped for that one letter and how large is the region on the magnetic disk where it will be recorded?

3) *Learn more about computers.* Even if I knew the mechanical and electronic mechanism of a computer, I still can't use a computer as much more than a fancy typewriter. I answer e-mail and even originate some messages, but I only venture occasionally into the World Wide Web. The friends who instruct me in these matters speak acronyms that condense the explanations but disperse the transfer of information. I am also disheartened by having a colleague refer to my three-year-old computer as a "dinosaur." Perhaps because of its age, or my age, I seem to waste a lot of time and paper reading the screen and then printing out messages for later pondering. Clearly, with less than two years to go before the millennium, I must master this universal tool and bottomless source of unedited knowledge.

4) *Learn more about how students learn.* I don't really think there's much new to be learned on this subject. It's more a matter of remembering. If we have ever really listened to our own children, we know about Piaget. If we have ever read explanations about phenomena that our students have written, then the results of the Force Concept Inventory came as no surprise. If we have ever had the privilege of sitting in the back of a lecture room while a colleague derived equations on the blackboard or talked about transparencies projected in a darkened hall, then we know the need for interactive lectures. We need to be reminded, though. It's easy to backslide.

You know what to do with new years' resolutions. Write them out and file them. Perhaps you will want to look at the file every so often as the school year progresses. As for myself, I will do such things—What they are, yet I know not. But I have a whole new year to work that out.

September 1998

I n a previous editorial (November 1995) I suggested that when faced with a physics problem, we should first write down the answer. It should be a reasonable first approximation, complete with units and error range. The answer will then help us shape our method for detailed solution of the problem, if indeed a detailed solution is necessary.

But first, of course, there must be a question. This prerequisite applies to more than just problem exercises. Ideally, every physics lesson, every lecture, every lab should be built around a question. The teacher should use the question as framework and should make the question explicit for the student. Every day on the left side of the chalkboard we should write down The Question. If all goes well, most of our students will learn a reasonable answer before the period is over. At least, they will realize that there is a question, and that's a step forward.

It might be instructive for us, the teachers, to begin by asking our students for first-approximation answers to The Question. Their answers might determine our strategy. Perhaps the lesson is unnecessary, or more likely we must fall back a few steps.

Of course, not all questions are good questions. Questions requiring memorization of formulas or definitions are usually bad. If the question can be answered by a multiple-choice exam, it's probably trivial.

Some questions are just plain wrong, or at least nonproductive. Generations of philosophers worried over the question "What keeps an arrow going?" The useful question is, "Why does an arrow stop?" A question appropriate for a physics graduate student might not be useful for a high-school student. For instance, "What are the Fourier transforms of common energy distributions?"

Questions should be raised in any kind of learning situation—science museums, for example. I have visited several this past year, all filled with ingenious demonstrations. Children of many ages were playing with the devices, enjoying the surprising phenomena. They pressed buttons, occasionally read the directions, and then moved on to the next exhibit. It was good entertainment for a rainy afternoon. There should be more to it than just entertainment, however. There ought to be learning, and for that purpose there ought to be questions, big ones and little ones. Perhaps the big ones could be major themes: Is energy really conserved? How can you generate electricity? Each exhibit might then raise a small piece of the question. Where does the energy go when water is heated at 100°C? Can you be killed by 10,000 volts? 100 volts? 1 volt? The questions should be prominently displayed, hard to avoid. Maybe there should be a giant question mark over the front entrance of each science museum.

Questions demand answers. If a class or a museum visit raises questions and doesn't require accountability, most people will take the easy way out. When the answers to pop quizzes in the newspapers are printed upside down, most people will turn the page around. Why sweat it? In classes, of course, we have exams. These should be tools of instruction, not just grading devices. In preparation for an exam, the teacher should review the questions that have been raised and then solicit student answers. *Reviews produce new views.* It's harder to quiz people in a museum setting. They aren't captive and their academic future is not at stake. They can be praised, however, or bribed with diplomas or medals. Champion museum goer of the week!

It takes skill and experience to ask good questions. The question must be important and should have ramifications beyond itself. The framing of the question should be intriguing, perhaps paradoxical. We should know the nature of most potential answers before we write the question, and yet we should be prepared for off-the-wall answers that may be better than we expected.

In his long life, Rabi learned the answers to lots of questions about the microstructure of matter. He used to tell the story, however, of when he was a boy and his mother would say to him when he came home from school, "Did you ask any good questions today?" She knew what makes good education.

May 1998

A while back I was sent a description of a foolproof way for students to solve physics problems. Each type of problem had a special solution guide. Step by step the guide required you to fill in blanks, leading you by logical sequence to the final answer. The teacher reported phenomenal success for his students when they used this system on exams.

Of course my immediate reaction was to recall the three-category classification system of Bacon's Scientific Method, or the sevenfold steps of Critical Thinking (analysis, application, interpretation, logical argumentation, organization, reflection, and synthesis). There is a similar system for successful mountain climbing, with the fifth item being "now climb your mountain."

The particular physics problem illustrated by a sample of this guided system concerned the position and speed of an object in free fall. Students were cautioned to fill in each blank with symbols, waiting until the final formula was derived before substituting numbers. Too bad. It turns out that the final speed was greater than the speed of sound. A simple numerical approximation at the beginning would have saved all that filling out of blank spaces.

Here's my suggested scientific method for solving physics problems. First, just write down the answer. How long is it going to take a ball to fall 10 meters? Several seconds, I suppose. Probably longer than 1 second, and certainly shorter than 5 seconds. That's a good approximation to an answer, complete with units. Time is involved; seconds are specified. Do I have to worry about air friction? If the time of fall is 1 second, the final speed is 10 m/s. That's about 22 mph (you have to know a few things with this system), and that's a lot less than terminal velocity for a baseball, but not a Ping-Pong ball. Besides, if the final speed is only 10 m/s, the ball couldn't travel 10 meters in one second starting from rest. So the time must be greater than 1 second. The Ping-Pong ball would take even longer. If the time is 5 seconds, then without air friction the ball would be traveling at over 100 mph. That's almost as fast as Sampras hits a tennis ball, and we know that's getting into the terminal speed range. So what kind of ball are we talking about here? And how precisely do you want to know, anyway? Or did you just want me to memorize $y = v_0t + \frac{1}{2}at^2$?

What's the wavelength of visible light? Since I can see a mm, the wavelength of light must be less than 10^{-3} m. It must be greater than 10^{-10} m, since I can't see an atom with visible light. I'll write down the answer as the geometric mean: between 10^{-6} and 10^{-7} m. Do you want a second opinion? Well then, the wavelength of light must have something to do with the magnification limit of ordinary microscopes. Did you ever see a visible light microscope with magnification greater than 10^3? With magnification of 10^3 I can identify details down to 10^{-6} m, since they would appear to me to be 10^{-3} m, which I can see. So visible light must have a wavelength a little less than 10^{-6} m.

But how do we know the size of an atom? Well, there are a few things that everyone ought to know. But if you want to calculate it, recall that the volume of a mole of solid is about 8 cm³, which has a cube root of 2 cm. A mole, as everyone knows, is 6×10^{23}, which is nearly 10^{24}, which has a cube root of 10^8. So there are 10^8 atoms lined up shoulder to shoulder along 2 cm. Evidently the radius of an atom is 10^{-8} cm, give or take a factor of 2.

What's the size of molecular binding energy? Every electrochemical cell yields 1 to 2 volts. Since each electron is shoved up a potential hill of a volt or so by the dissociation of a molecule, the binding energy must be about an electron volt. Since light in the near ultraviolet can dissociate molecules—tanning, bleaching, photography—the energy of a photon of light must be 1 or 2 eV. If there is some reason to know more closely, you can look up the exact values of Planck's constant, which, of course, isn't all that exact.

This approach to physics problems is the same as we use in everyday life. If you go to the store for a new coat, you figure out in advance the price that you will pay, probably to within 25%. You don't calculate the price to four significant figures unless there is some very great and definite reason. (The sales tax will defeat you.) That original estimate determines where you will shop and what you will look for.

Working on a physics problem should be like buying a coat. What are you looking for—a velocity, a mass, a time? Within a factor of 2 or 10, what would be a reasonable value? Write it down, including the units. Then ask yourself, what precision do I need in this problem? Maybe an order of magnitude answer is all that is useful. If there is some good reason for getting more precision, at least you have a starting point and a reasonable answer to check against when you get a next approximation. Furthermore, the first answer is probably your best guide to an appropriate method for making the next approximation. Circle around the answer. What relationships does it have to other things you know? (For instance, remember the relationship of the energy of visible light photons to molecular binding energy. Remember the relationship between final speed and distance traveled.)

Of course students aren't used to this method of solving problems. If you require of them a fixed approach to each type of problem, and three meaningless significant figures, they'll probably learn how to do it, but they'll get the wrong impression about real physics. After all those years of math and science courses taught in the traditional manner, they probably would prefer to operate in the conventional mode. It will be hard to wean them, but physics courses ought to be an introduction to the way the world really works. Use this method yourself when you show a class how to work out a problem. Start out every time by bullying the class into helping you write down the answer first. Once you've got the answer, the solution should be simple.

November 1995

Mea Culpa—et Tu?

When Einstein first formulated his great equation of general relativity, he realized that it predicted a continual expansion of the universe. In order to correct such a nonsensical result, he inserted a "cosmological constant" into the equation, thus ruining its symmetry but making the results more plausible. As Einstein said later, after it was discovered that the universe was indeed expanding, "Adding that constant was the greatest mistake I ever made."

I know how he felt because I have made mistakes too. Usually mine are not so profound. Years ago I wrote a high-school physics text that made use of a second color for graphs and phrases that needed emphasis. On one page there's a derivation that step by step leads down to a remarkable result at the bottom of the page. "Oh, boy," I told the reader, "Look where all this leads!" Nothing. Blank. The second color never made it to that page.

There was a standard introductory quantum mechanics text that everyone used a generation ago. Of course there were mistakes in it— typos, wrong answers, etc. All books have these. Legend has it that the author printed form postcards that said, in essence, "Congratulations for finding the mistake in my book. Thanks for telling me about it, and rest assured that in the next edition I will leave it just the way it is so others may have the same pleasure that you have had."

Mistakes like that can be very educational. I once proposed to a publisher that we create an introductory text to be called *Phallacious Physics*. We would guarantee that there would be at least one egregious mistake per page. The reader would have to be alert every minute.

Because of my own failings, I was heartened to read the other day that the Post Office has spent 1.2 million dollars on a new stamp series featuring Wild West characters, including the foremost black cowboy, Bill Pickett. It's an appropriate honor, all right, but they apparently portrayed his brother, Ben, instead of Bill. Mistakes will happen, I always say.

Consider, for instance, the graph of amplitude versus frequency for the harmonics of the square wave in the December centerfold. The derivation in the tutorial goes to some trouble to point out that the even harmonics are zero, and that the three harmonics calculated are for $n = 1, 3, 5$. So if you will kindly use a green pen and some white-out, you can easily correct my mistake on the graph by moving those bars from 1, 2, 3 to 1, 3, 5, and I would be grateful to you. (How could that have happened? I must have seen that graph a dozen times before it was printed.)

I'm sure, Dear Reader, that you yourself have never made a mistake in physics. You're the only one, you know, and there's a caution you should be aware of. They say that Indian rug weavers always make one deliberate mistake in each rug, lest the gods be jealous. A word to the wise, I always say.

March 1994

What the Student Saw

The day before the first exam a group of us were going over some vector problems. I proposed that a balloon had traveled 4 km east, 2 km north, and 1 km up. "Wait a minute," said one student. "Isn't north up?"

Well, of course, on my blackboard, north is almost always up. I don't see it as up, but the student did. Meanwhile for weeks I had been writing and drawing symbols and diagrams far more arcane than map directions. Who knows what the students had been seeing?

When you or I see dx/dt, we see a slope on a graph, or maybe a car racing by. When someone writes $\int v\, dt$, we see the area under a curve, or maybe a distance traveled. Do you suppose our students wonder if the d in the derivative stands for distance and think of the integral sign only as a peculiar formula to be memorized?

What do my students see when I draw arrows on the board to represent forces? In his little book on vectors, Banesh Hoffman suggested that the French missionaries in Canada brought with them their Cartesian math as well as their religion. The missionaries persuaded Indians that vectors were arrows and arrows were vectors. One tribe became so converted that henceforth they sent their braves out in pairs to shoot arrow-vectors in perpendicular directions, thus to bring down the fallow deer with the powerful resultant. To what image of reality have we indoctrinated *our* students?

I'd like to think that our physics models and symbols, and the way I portray them, are all self-evident and obvious. After all these years, they certainly are to me! A slanted line on the board is clearly an inclined plane, unless it happens to be a plot of $F(x)$ for a spring, or is it current as a function of voltage? If there's any question, just look at the fine print on the axes.

If students can't grasp our symbols immediately, or if they don't notice the fine print, perhaps it's because they are stupid or wicked. I might be tempted to think so except for one thing. I've been playing a trumpet or have been singing in choirs for most of my life. By now I'm reasonably competent in figuring out one note at a time from the printed score. Sometimes I can even read ahead a few notes and press the right combination of three keys or pitch my voice in the general range. But our choir accompanist can instantly read four or more notes at once, making all her fingers and even her feet do the right things. She can even sing the words at the same time. How does she do that? What does she see that I don't?

Consider how clever any of us are to see meaning in ordinary words. *Love.* You didn't rhyme it with *hove* or *move.* You rhymed it with *of,* which for some reason ends in a *v* sound. How did we learn to see such complex symbolism? Indeed, the symbolism is deeply layered. With the word *love* do you picture family warmth, baby smiles, abstract goodwill, joyful lust?

When we're trying to convey our mental pictures of reality to others, whether with words or diagrams, we should use many modes and many models. When possible, blackboard diagrams should be in close proximity to three-dimensional scale models of the same thing. The math symbols should be presented in parallel with diagrams, and the diagrams accompanied by words and quantitative examples. Of course our numbers should always be in terms of units, but beyond that, the size of each number should be interpreted in terms of familiar quantities, if possible in terms of feelings in our muscles.

One picture is worth a thousand words only if the picture is recognizable. Seeing is not necessarily believing; sometimes, seeing isn't even seeing.

January 1991

Why Do We Need Labs?

We got a letter from a teacher in New York State, of all places, saying that his school superintendent was proposing to do away with the extra double period for physics lab. The resulting released time in that particular high school could save half a teaching position. As an alternative, the principal proposed that time spent in teaching lab might be equated with corridor patrol duty. He apparently figured that you're on your feet in both cases, and, of course, preparation time is equal.

Better yet, why not do away with physics labs entirely? Board knows, they're expensive. Is there any education research proving that labs are worth the trouble? Do students who spend time in laboratory do better on final exams? Not that I know of. Imagine the difficulty of doing a meaningful education experiment on this question. What kind of exam should be used? Who teaches the control group? Is time equivalent to the lab period spent on test problems? Besides, what type of lab should we test?

Some years ago Tom Miner played Devil's Advocate and drew up a list of good reasons for not having physics laboratories. If you're looking for excuses to drop lab teaching, just clip these out and send them to your nearest school board. On the other hand, if you, too, are afraid that your administration is seriously thinking of doing away with labs, burn this page. (n.b. these arguments apply only to high schools. Everyone *knows* how effective labs are in college.)

Educational reasons for not having student labs
1. Demonstrations by the teacher are more effective and are more likely to produce correct results.
2. Students frequently get incorrect results, and learn to lie, or round off.
3. Student lab reports are invariably neater and better organized if copied from the board.
4. While performing the lab-demonstration, the teacher can point out special features of the apparatus and stress important points.
5. Since the teacher is experienced with the apparatus, time-consuming pitfalls can be avoided, leaving more time for teaching.

6. More sophisticated experiments can be observed on film, performed by experts, using apparatus not available in schools.

7. Laboratories are often used in the morning for homerooms, creating a forbidding environment for students who are not science-prone. If labs were converted to classrooms, there could be a friendlier milieu for everyone.

8. Scheduling a double period for lab is troublesome for computers or assistant principals. Without this complication, students would be able to sign up for more cultural, relevant, and self-fulfilling courses.

9. With labs held only once a week during the double period, the laboratory proof of laws frequently gets out of synch with the classroom presentation. If all such proofs were done by the teacher, the students would learn each step in the correct sequence.

Financial reasons for not having student labs

1. Abandoning a double lab period would immediately reduce teacher work load by 16.667%. For instance, the teacher could then do six sections instead of five, or could be given cafeteria assignment.

2. Once the equipment is moved out, many lab rooms would be big enough to be partitioned into two classrooms. Alternatively, the large room could be used for team teaching.

3. With student labs, it is usually necessary to have twelve sets of each apparatus. With the teacher doing the lab, only one set of apparatus would be needed. The annual equipment budget for physics could then be cut to $1000.

4. Without the needless duplication of apparatus, equipment storage cabinets could be used for forms, films, and coffee supplies.

5. The time now spent on repair and maintenance of apparatus would be practically eliminated. The laboratory assistant could be reassigned to the bus pool.

There you have it, friends. I was going to present my own arguments in favor of student laboratories, but Tom's list has overwhelmed me and I've lost heart. If you can think of any good reasons for having labs, let me know so that we can carry a sequel. In the meantime, I'll be looking in the Yellow Pages for buyers of old laboratory equipment.

January 1979

All Out For the Physics Lab!

In the January editorial, with tongue in cheek, I listed many reasons for abandoning labs in high school physics. You've been sending many contrary reasons, and, of course, I have a list of my own. We'll put those all together next month. Meanwhile, let's consider the nature and purpose of labs in introductory physics. What is this special feature of physics instruction?

First of all, labs ought to be fun. Now, there's all sorts of fun to be had in this world, and it's unlikely that physics labs will ever compete with a good party. But there's no reason why physics labs shouldn't be just as popular as band practice, or art class, or gym. In all these cases there is a break in the usual school routine of sitting at a desk. Students can move around, talk with each other, and handle something besides books.

Usually you have more fun doing something if you expect to have fun. There has to be some expectation and some promise that the experience will be pleasant and maybe even exciting. A little hype will help. Before they get to lab, students ought to be given some forecast of what they will see or do. Perhaps in the previous class they can see some brief action with the apparatus or face some paradox to be resolved the next day. Not all labs should march toward known conclusions. Occasionally there should be mystery unresolved, or complete surprise. Why should research physicists have all the fun? Even the addition of vectors can be lifted from boredom. Let the vectors march around your school building, or survey your neighborhood. Give your students a glimpse of the nontrivial nature of vectors by having them try the simple addition rules on the surface of a globe. A guiding principle for your attitude toward a lab exercise might be that the hype should be genuine. If you yourself don't look forward to a lab, change it or abandon it.

But don't the students need the drill, even if you don't? Sure, students have to play scales in order to learn to play piano. But not all the time. Jazz it up. Is lab the place to teach neatness? If you think so, pray your teenagers never see a real research lab. At any rate, neatness to the point of fussiness is foreign to our free wheeling traditions. Must the students learn precision? Far better that they glimpse the real world of having to judge when precision and the time needed for it is justifiable.

In the introductory course, neither high precision nor great accuracy is usually justified.

Is laboratory the place to *prove* nature's laws? Who are we kidding? In our labs, energy is always lost, objects never fall freely, planes always have friction. The profound generalizations can be illustrated, but not proved. Can our students at least *discover* laws in laboratory? Inquiry teaching? That silly illusion! How presumptuous of anyone to think that in 50 minutes (less the time for setup and putting away), a student can rediscover the great discoveries. (Now, Dr. Hertz, this afternoon we will discover that light is electromagnetic radiation. Be sure to fill neatly all the blanks in your workbook.)

What's lab for, if not to discover and prove? Why, it's the place to *handle* phenomena. Most of our students have never experienced the most common processes of their daily lives. They have never observed a ball rise and fall. They don't know which way the wind blows. They have never really looked into a mirror. They've never taken apart a battery, or a gear system, or a camera. They shouldn't just be shown these things, particularly with book pictures or film. They should get their hands on them. Let them play with the variables. The play may be guided, but only in part. After a student handles the phenomena, she may be ready for more formal analysis, and after that, back to the lab to see the phenomena with new insights.

Can there be no research in labs, no Eurekas? Sure, there ought to be small Eurekas every day. But real research takes time. For a genuine research experience, there must be a need to know, feasibility studies, tinkering, and blundering, and false starts. Finally there must be reasonable reporting on what was done. Every physics student should experience such research, but not weekly. There's time for only one or two such projects per year.

In a later editorial, we'll analyze some standard lab exercises and see how they meet the criteria we've set up. In the meantime, keep sending in those comments and denunciations. It turns out that not everyone agrees with our revealed truth.

March 1979

The Varieties of Laboratory Experience

I n this special issue devoted to high school physics laboratory manuals, it's appropriate to raise questions about the nature of the lab. What sort of experience does it provide students? Well, if "physics" is defined to be what physicists do, then school physics lab must be what the lab manuals tell the students to do. All of these manuals tell the students to set up apparatus, make measurements, and graph data. That's what physics lab is.

Now that we have defined the nature of the lab, what is its value in the curriculum? Labs are expensive in student time, faculty preparation and grading, administrative scheduling, school space, and apparatus budget. As far as I know, no one has ever demonstrated (in properly controlled tests) the value of the laboratory experience in terms of better grades, higher enrollment, or better preparation for future courses. Then, why have labs?

We all know, deep in our bones, that it's good to have labs, because we ourselves had them. Besides, there are all these lab manuals. Some people think that only in the laboratory can students learn the "scientific method." Note the prevalence of lab reports that have the standard format: purpose, apparatus, method, data, comparison with theory, explanation of errors. A clever student, with conveniently arranged blanks in the lab manual, can cover all these points in one hour, two at the most.

Then there's the "little scientist" or "Eureka" or "discovery" reason. You learn about science by being a scientist. Instead of just reading about a physics law, the students are guided to discover it for themselves. (Don't tell them. Instead, help them to realize that . . .) With the proper guidance, students can discover Newton's second law in 45 minutes. It's a pity that Galileo didn't catch on that fast.

Another reason for having lab is to prove physics laws so that students won't have to believe their teachers or textbooks without further substantiation. Evidently, this theory is based on the assumption that these sources are unreliable, a libel that I do not wish to pursue further. There is some point, of course, in teaching science students to be skeptical of authority and here the laboratory comes into its own. As we all know, energy and momentum are never quite conserved in the laboratory, proportional data when plotted do not line up with the origin, and with student-wired circuits, I = 0 or 20 amps instead of V/R. Fortunately we deal only with bright students in physics, and they know how to discover and prove the things we tell them, regardless of what they observe. If not, in their reports they will ascribe the discrepancy to "human error." And Lord knows that's true.

I think that there can be some valuable outcomes of the laboratory. First, most students need experience in using tools, even the screwdriver. They could get this in shop, of course, but many physics students take Latin instead, and nice girls don't take shop at all. Unfortunately, if our lab apparatus is too well made, and is all set up for efficient use, the kids won't have to use screwdrivers. Maybe we should design broken apparatus and furnish sets of tools.

The second skill that lab could provide is how to do a real experiment. Since real experiments cannot be done in 45 minutes, or even in two such periods, perhaps we could require one experiment per semester. These could be done at home. To make it real, each student should prepare a feasibility paper before beginning: what's to be done, what equipment is needed, what precision will be attempted, how long it will take. (Most preliminary feasibility studies will describe unrealistic projects.) While the work is being done, real log books should be kept, and at the end a real report should be written.

A third valuable activity in lab is simply handling the phenomena. Students should personally feel the forces of springs and magnets, should reflect and refract and diffract light, and should be shocked by static charges. Handling these phenomena is very different from watching a teacher demonstration, or worse yet a video film, or much worse yet, a computer simulation. The students themselves should play with the apparatus, learning how to operate it, and taking simple data. This "playing" of course should be subject to testing, like all their other work.

The good thing about teaching lab is that you are your own expert in knowing how to do it and why you do it. With 4,000 full-time high-school physics teachers in the U.S. there are 4,000 different types of lab teaching. That's where our artistry comes to the fore. Music teachers may be noted for their concerts, and English teachers may be treasured for directing the senior play. Physics teachers are remembered for the afternoon everyone timed a ball dropping from the school roof, or the day the water spilled from the ripple tanks. That's physics.

April 1984

L arge lecture courses usually have a bad reputation. Everyone knows stories of classes where 500 students are supposed to listen to a dry lecturer two or three times a week and then attend a lab taught by an uninterested graduate student. Such situations are thought to be common in our large universities. Far be it from me to deny that such situations occur. I can cite my own collection of similar horror stories.

However, the corollary that lectures are *ipso facto* bad, is not at all true. Nor is it the case that large lectures are necessarily bad but small lectures are always good. First, note that most classes of any size are mostly lecture classes. The geometry of the classroom is designed for students to sit and listen, and for the teacher to stand and talk. Granted, there are some gifted teachers who can create true conversations or recitations with small groups. But I ask you, in most classes, most of the time, who does the talking?

Whether or not some magic boundary size exists between responsive class and faceless audience depends on the physical nature of the auditorium and the skill of the lecturer. I have seen a full professor mumbling at the blackboard, quite out of touch with his lecture class of three students. On the other hand, I have seen a grand-old-man reading a set-piece lecture from a podium and holding 200 students spellbound. We have all seen, at least on television, that magnificent appeal by Martin Luther King to an audience of hundreds of thousands.

It is manifest that the lecture method is the most popular, the easiest, and probably the least expensive system of instruction. A niece of mine once sat through many lectures by a nationally known science educator, urging his class of future teachers to make their students get out of their seats and learn by doing. Haven't we all at our meetings heard lectures on individualized learning or computer assisted instruction? All you need for a cheap physics lecture is a captive audience, a speaker, and a blackboard. If the blackboard causes difficulty, the speaker can use a viewgraph. Dim the lights, and you can do without the audience.

Lecturing is an art that can be enhanced with training. There are common sense rules about how to write on blackboards (first, break the chalk in half); how big to make demonstration apparatus; how to have only one theme per lecture; how you should begin with what you're going to tell them, then tell them, then tell them what you've told them. Every lecturer should arrange to be videotaped at least once a year, in order to catch annoying idiosyncrasies. We really ought to have classes and American Association of Physics Teachers workshops on how to lecture. No doubt they would turn into lectures on how to lecture.

What should be the role of a lecture? Not to pour in information, at least not usually. There are better ways to provide facts; that's what textbooks are for. Not to prove laws of nature. Students believe them anyway. Not to solve homework problems. The students must do those for themselves.

What's left? Why, that old fashioned word, inspiration. Or, if you prefer, the demonstration and embodiment of an attitude. Is the point of doing a lecture demonstration to show a phenomenon? Only in part. The more important demonstration is to show how a skilled scientist manipulates apparatus. Beyond that, the most lasting effect is produced by the visible excitement and pleasure of the demonstrator. No textbook or movie or computer console can substitute for the genuine enthusiasm of a good lecture demonstrator.

But don't you have to use facts and impart information as you lecture? Sure, but you can do so with emphasis and style and body language. "Why should F be proportional to a when, as you can see on the scale, I am dragging this wagon with constant force at constant speed?" That sort of challenge can be done in a text, but not in the same memorable way. "And now let's put some numbers in this equation and see if the result makes physical sense. Let's see, 10 divided by π is 3, sin 10° is 1/6, since sin θ is about equal to θ for small θ; naturally we're going to get an order-of-magnitude feel for this quantity before turning to our calculators. Aren't we? You bet we are, because that's the way grownups do things."

Maybe these points don't reflect your particular style. But you can be sure that the main thing you're teaching *is* style and attitude and enthusiasm—or the lack thereof. The facts, and in our case the laws of physics, we teach by text and exams, or threat thereof. Some people are stage hams; the bigger the audience, the more the interaction. I have had some great teachers of this type, teachers who influenced my life. Some people can only work with small groups of students or with one at a time. I have known and been influenced by teachers of this type also.

A good institution arranges the setting to match the teacher. No geometry of classroom, or scheme of organization, or system of administration is of ultimate importance in the teaching-learning process. All that matters is that the system allow contact between a student and a knowledgeable, enthusiastic teacher. That can happen with a student at one end of a log and the right person at the other end. It can also happen with a student in a 500-seat auditorium and the right teacher up front marveling at a phenomenon that has just been produced and analyzed. Little lecture or big lecture; it really makes no difference. As with any other mechanism of education, all you really need is a good, and experienced, and devoted teacher.

October 1983

How to Write for Either *TPT* or *Playboy*

S o somebody says, "You should really write it up and send it in to them." Publication may bring you wealth (though precious little from *The Physics Teacher* [*TPT*]) or fame (very fleeting). Before you sit down with pen, or typewriter, or word processor, there is one thing you must know. It is as important as knowing your subject. You must know your audience.

You have to keep in mind an image of that audience all the time you're writing. Preferably, you should picture a particular person. You must decide whether your reader is young or old, at home or at work. You should know which words will be familiar to that person and which must be explained. You must know how the reader can enjoy or use your information.

One way to study the reader is to study the magazine. Presumably the subscribers have been reading the previous issues. At least the previous issues reflect the editor's image of the readers. When it comes to getting your manuscript accepted, it may be more important to match the editor's image of the reader than your own.

In the case of *The Physics Teacher,* a casual inspection of the magazine will indicate that the editor holds the readers in high regard. He assumes that they like teaching but may well grumble among themselves about students or administrators. He knows that they like physics and would pursue it as a hobby even if it weren't their business. A considerable range of technical background is assumed, with the confidence that lots of readers will take the trouble to stretch occasionally, perhaps because they have confidence in the editor.

Here are some points you should notice about *The Physics Teacher.* There are mechanical constraints. In each issue there are three or four major articles, about 5000 words each, copiously illustrated. Before writing such a major article, you should proposition the editor with an outline. If there is tentative agreement and you start to write, consider the next hurdle. Even if the article were accepted, would it be read? Picture the magazine on the coffee table in somebody's home and imagine that there's only a few minutes before bedtime. Or perhaps it's on the office desk, and it's at the end of a long day. The subscriber picks it up and idly thumbs through to your first page. What's your grabber? Can you capture that physics teacher in your first paragraph? Will your article be useful in teaching tomorrow's lessons? According to surveys, our track record in this regard is very good, and we intend to keep it up.

The Physics Teacher also contains many shorter contributions in the Notes section. These should be under 1000 words, and usually no prior discussion with the editor is necessary before sending in the manuscript. However, the same criteria for acceptance apply to the Notes. So do the mechanical requirements listed in the February 1984 issue: double-spaced manuscript with generous borders, two copies, inked drawings, glossy photos, references, etc. We need two copies, easy to read, because every published manuscript gets read at least seven times before it's printed.

What happens when someone sends us a single copy of a 2000-word note, single spaced by a word processor with dot-matrix printing? Well, it makes us short-tempered with our students that day, and we'll probably send the paper back after wasting too much time trying to see if it can be saved.

What happens when we get a manuscript with an abstract on the first page, or with lots of differential equations in it? We conclude that the author has never read *The Physics Teacher* and therefore probably doesn't know the audience. We'll read the manuscript but we're already prejudiced.

What happens if the manuscript is laced with jargon, either from physics or from science education? We will take a dim view of the paper, translating it only if it appears that a gem may be hidden.

Before sending your manuscript off to the notoriously critical and ill-humored editor, try it out locally. If you teach in a high school, try to get a local university or research physicist to read it; if you are the latter, seek out the former. Every writer needs an editor, preferably before the real editor sees the finished product.

The title suggests that these rules are general and apply to all kinds of writing. Indeed, suppose you want to write about quantum chromodynamics. Consider how different your article would be in *The Physics Teacher* than in *Physical Review Letters.* It would be longer for one thing, with far fewer authors. How you would write it for *Playboy* boggles the imagination, but it would need better photos, I should think.

October 1984

"Listen," my colleague said. "I want to know what you think about this." For the next twenty minutes I patiently listened, learning more than I ever wanted to know about a problem my friend had. I hope it did him some good to tell me all that, but he never did hear my opinion. He did all the talking.

A one-sided conversation is a common experience. I suppose I'm as guilty as the next party in turning a conversation into a monologue. It's an occupational hazard. Kid asks a question; I give the answer. At length. After all, that's why the students sit in rows facing the teacher.

And when I ask a question of the students, I want an answer right away. Perhaps I wait a second or two. Several decades ago, Mary Budd Rowe measured the intervals between teacher questions and the next words spoken by the teacher. The average time was a little over one second. It was embarrassing. Very few students can leap into a conversation with only a second or two of preparation. It turns out that if you wait at least three seconds, the student will frequently have something worth saying.

Here's an interesting project for some aspiring social scientist or Ed major. Measure the ratio of class time during which students talk to the time the teacher talks. In a straight lecture, the ratio is about zero (except for the whispering going on in the back row). What is it in your classroom, or recitation section, or lab?

Of course, there's the danger that if you listen to the student, it will take up class time and you will hear nonsense anyway. There's a legend of a professor in Germany a hundred years ago who gave very formal lectures. One day a student raised his hand in class. When the bewildered professor paused, the student asked a question. After pacing awhile, the professor responded, "If you persist in asking these questions, we will never finish the work of the course." I know how he felt. When I was fresh out of grad school I ran a Physical Science Study Committee (PSSC) workshop for physics teachers on Long Island. One teacher would frequently interrupt my lucid lectures to ask a question. Those questions were so stupid that I often didn't know the answers. It took me awhile to realize that this man (who was a great teacher) was asking the questions that everyone else was too confused or too timid to ask. In the March 1997 issue there's a note about the physics of falling coffee filters. The story has a moral that's more important than the equations for terminal speed. The authors gave an open-ended lab assignment with no cookbook instructions to a group of bright freshmen. It was a disaster, but a revelation. When you listen to students you discover that they don't perceive things the way you do. You then have to face the problem of finishing the course or trying to bring students up to t_0. The decision may be that the standard minimum material must be covered, letting the students fall where they may. Or you may undertake rescue measures. But if you don't listen, you won't even know there's a problem.

Of course we listen to the students every time we give an exam. However, if the exam is multiple choice, or even routine homework type problems, we may not be hearing much. These exams are easy to grade but hard to interpret. I once took over a sick colleague's course and administered one of his multiple-choice exams. It was his custom to allow the students to justify their choices if they wished to, in hope of getting partial credit. There were more cases of students choosing correct answers for wrong reasons than the other way around. Now I never give an exam where there is not the opportunity for the student to explain or describe something in English sentences. These are as easy to grade as standard problems, and far more revealing.

The next time you get cornered by somebody talking nonstop, use the time profitably by considering how it is when the tables are turned and you are conversing with students. Whenever you are tempted to deliver a full lecture in answer to a simple question, keep in mind the immortal words of Ring Lardner in his story, *The Young Immigrants.* "Are you lost, daddy?" I asked tenderly. "Shut up," he explained.

March 1997

t's hard to write a September editorial in July, particularly when you are in the middle of winter. I've been attending the biennial conference of the New Zealand Institute of Physics. Everything is just a little different down here, including the seasons. Actually, I am getting used to looking to the right before crossing streets and have begun thinking that folks at home must be down under. Evidently, here we are up over, but the moon is upside down.

In such an upset state of mind, I have been wondering how I would teach physics and what I would teach in a completely new world. What would you do if you were familiar with the concepts and methods of physics but had never before seen an introductory text? Would you necessarily start out with vectors and kinematics, etc., etc.? Would you rely on wave optics to explain interference or would you rather appeal to extremum paths of photons?

Once, long ago, I worked out an introductory approach to thermo where temperature was directly related to average communicable energy of the microstructure. I proposed using these ideas in a book, but Mark Zemansky talked me out of it. He had two reasons: first, he said, temperature is a primitive concept involved with touch and should remain fundamental; second, if I adopted this valid but novel approach, it would be unfamiliar and no one would buy the text. I knew a philosophical argument when I heard one, and started the heat chapter with the old demonstration of three pans of water—hot, cold, and room temperature.

So, where would you start if you had no prior constraints and didn't want to make any money? Let me know, and we'll publish your suggestions. Meanwhile, here are my ideas. Since very few high-school students of physics will become physicists, there is no need to specialize in training them for such a career. The traditional high-school course is just a scaled down version of the introductory college course. According to Sadler and Tai in the May 1997 issue of *The Physics Teacher,* success in college introductory physics is only weakly related to whether the student took high-school physics. Since it appears that a preliminary run-through of the standard course doesn't help much, and since many students will not take college physics anyway, why not free up the high-school course? For that matter, why not free up most college introductory courses, except perhaps for those few students who will go marching through the four-year sequence?

In my upside-down world, I would start out by scaling the universe, in length and time and energy. You might think that energy is a sophisticated concept, but it can be made tangible and primary with many homely demonstrations. In the process of relating the microworld and the cosmos, it would be vital to examine measurement methods at every level. However, vectors would not be necessary, nor would it be required to know the difference between distance and displacement, or speed and velocity.

It would, however, be necessary to compare sizes. Everything should be quantitative, with every numerical value interpreted in terms of some familiar unit. There should be continual use of order-of-magnitude calculations.

While my primary theme would be the nature and scale of the universe, there would be many opportunities to apply the concepts and quantitative techniques to everyday objects. How do things work? Dick Crane has been helping us learn such things for years, and now Louis Bloomfield at the University of Virginia has made a popular course out of such questions.

In linking the micro and the macro worlds, we would have to learn about measurements requiring quantum ideas and reference frames. We surely would avoid any rote memorization of particle schemes, but would investigate phenomena requiring models of the microworld. Part of these models consists of the four interactions between particles and their everyday manifestations. Gravity and static electromagnetism can be linked by analysis of potential wells and examples of force as a function of separation distance. Consideration of reference frames leads into velocity-dependent forces and currents. The strong and weak interactions explain radioactivity and energy production in stars. In considering interactions between systems, conserved quantities would play a prime role. In this view, force becomes a derived function equal to

$$\frac{\mathrm{d}p}{\mathrm{d}t} \;or\; -\,\frac{\mathrm{d}U}{\mathrm{d}t}\,.$$

A proper physics course should celebrate the profound insights that have been developed in the last few hundred years. It should teach the quantitative skills that allow students to understand and participate in those insights. While glorying in what we do know, throughout the course there should also be a sense of mystery and wonder. Students should discover how little we know in some areas and how fragile is the human enterprise on this thin shell of Earth.

Perhaps here, at the antipodes, my thoughts are upside down, but even where the moon is right side up, wouldn't it be fun to develop a brand new way of looking at physics?

September 1997

The Variety of Learning Experiences

onsider now the *Physics Learner.* We usually meet this person in a course for which we prescribe a sequential series of lessons. We choose a text, organize laboratory exercises, assign homework problems, and administer frequent tests. The heart of our service is the lecture. There we tell the *Learner* what is important in the text, thus providing clues about the topics to be examined in the test. Since the *Learner* is unfamiliar with the subject, there is no need to base instruction on what the *Learner* already knows or wishes to know. This state of affairs is illustrated by a picture of the head into which knowledge is being poured. Unfortunately, it is all leaking out the bottom.

The situation isn't really that bad. Throughout school days students are immersed in a system where the facts are poured in. Most of these facts are forgotten, which is just as well, but as the students age, more and more facts get used and assembled in the brain or fingertips. The high-school senior is clearly more capable, at least physically, than the seventh grader. Schooling does seem to help most students, even when it gives the illusion of squelching enthusiasm for learning.

Suppose we didn't know about our standard and partially successful system of education. How would we organize the effort? Why not first ask about the *Learner,* and figure out all the ways we know about how people learn things? Consider your own experiences, or those of students you have known.

People learn many things by rote. In some cultures, memorization by repetition of a few classical texts is the only formal education. This kind of instruction has a bad reputation in Western culture, but when the memorized information is important it is a very effective method. Clearly there is great practical usefulness in being able to recite the alphabet or to know the multiplication table. There is no better way to learn these than by rote repetition, preferably in childhood. Even in adulthood, musicians learn to sing or play by repetition. Somehow the fingers or the throat learn during the first run-through so that the second time is easier. Physics teachers gain expertise in solving simple physics problems by solving them over and over again, year after year. With enough practice, the method looks like the logic of an expert.

Some people learn best studying by themselves, reading texts and solving problems. Others do better in social settings, teaming up with one or two others and explaining things to each other. The problem in organizing a learning environment is to cater to the needs of each and to the pace of each. Many attempts have been made to provide individualized instruction, either by paper workbooks or with computer programs. Some students like these systems, but there have been no outstanding successes, at least none that have survived and propagated. Some people should study alone; some do better in small groups.

The common experience is that cyclic learning is useful. If at an early age you literally handle phenomena illustrating Newton's laws of dynamics, and then in high school describe the results algebraically, and in college run through the explanations in freshman year and again in junior year, then you are in a good position to begin to understand the laws when you first start to teach them in graduate school. The trouble is that only a few people specializing in physics will study the subject over and over again. We give most students a course in one or two semesters, and at the end lament about how little they understand.

In our rush some years ago to beat the Russians or the Japanese, many schools succumbed to the temptation to shove advanced material into lower grades. Thus we have college freshmen who have "taken" calculus but who can't do algebra or trig. We all know the silliness of teaching elementary-school children about atoms or genes. The Piagetian fad in education has faded, but you still can't teach a six-month-old child to walk, and you can't teach most elementary-school children to reason by analogy. In tempering the wind to the shorn learner, we must be aware of age limitations on concept development.

Even students who have achieved the formal operational stage of Piagetian development like to see concrete examples. Physics instruction without demonstrations is like dinner without food. To be sure, there are a few students who can understand and use information that has been introduced with only symbols and math derivations. But most learners need to see phenomena, preferably under their control. Better yet, they should feel it in their muscles. The bigger the muscles, the longer the memory.

What drives a *Learner* to study something? There has to be a need to know. Perhaps the need is to pass a test. That's a valid purpose and a useful goad. Hovering over every exam is an implied need to know. It is called fear. A student's future can be changed completely by an exam grade. If a premed fails the Medical College Admission Test (MCATs), there goes the country club membership. The main virtue of a formal course of study is the provision of exams at regular intervals. Of course, simply cramming for an exam usually leads only to short-term memory. On the other hand, reviews produce new views. A good teacher exploits reviews and exams more for teaching than for grading.

A more effective goad for the *Learner* is simple curiosity. To satisfy that, people would not normally take a course. Instead they would use acquired skills of library search or on-line computer references. The trouble with disorganized searching for understanding is that a lot of information must be approached sequentially. In college this problem is recognized by establishing prerequisites.

A good teacher can supply the seeds for curiosity and a need to know. That's the purpose of the lecture. Facts can be obtained from the text. In a lecture the teacher can wax enthusiastic about a deriva-tion, marvel over a demonstration, view a false explanation with scorn. For every answer, a skilled teacher points out new questions. A good teacher inspires; there is no better word for it.

The proof of learning is usually taken to be the results of those course exams. They allow the *Learner* to demonstrate skill in problem-solving and familiarity with many specialized terms. There are now attempts to judge students in terms of projects they have tackled. This so-called authentic testing is meant to determine if the student can integrate what has been learned and create something novel or useful with it. Like most educational fads this idea has a valid kernel, but it has proven devilishly hard to implement the idea in practice. We all have known students, or colleagues, who appear brilliant on exams, but as the years go by they never really accomplish much. And there is the inverse, the student who does only medium well on the multiple-choice exams, but somehow links disparate ideas into a new and useful discovery.

So how do people learn? By rote, and by logic. Singly or in groups. In lectures or in lab. Immediately or after many cycles. Within their con-ceptual range. Because they have a need to know. Because they are inspired by a good teacher and their curiosity is raised. And when they have learned enough, they may or may not be able to use it by creating something new.

If everyone learned the same way and at the same pace, there would be no difficulty in teaching. However, because learning takes place in so many ways, our profession is more an art than a science. If we try to impose any particular system of education we will harm some of our students and frustrate ourselves. Physicists have a reputation for free-wheeling attacks on problems and a healthy disregard for rigid rules. Because of the many differences in our students, we should avoid the shibboleths of fads and use many different teaching methods to provide a wide variety of learning experiences.

January 1997

What Every Young Student Should Know

ow . . . here's your average red-blooded American boy or girl, eighteen years old and a high school graduate. What does our school system guarantee that he or she can do after these twelve years of training? The English teacher guarantees that he can read the newspapers and write either a friendly or a business letter. The social studies man guarantees that our graduate recognizes the name of the president of the country and knows that July 4th is a holiday. The math teacher assures us that any graduate can add grocery bills, and most of the students will surely multiply. Our physical education colleagues warn us that 50% of the nation's youth are below the national average in strength tests, but we can still be assured that the high school graduate can rise to a sitting position, keeping his knees straight. And now we science teachers—what do we guarantee?

Come, come! Surely we can claim that the high school graduate can explain why a steel boat can float. Is it possible that in this day and age he will not know how to use a voltmeter? Of course, at some time in school every student is required to plant seeds and study their growth in a controlled way. Would not every high school graduate, these future citizens of a country that spends five billion dollars a year on space, know how to locate Mars in the night sky? At least they would know the moon's trajectory through the heavens. How about knowing when the sun rises? Or where?

Perhaps there is no need to be embarrassed. After all, science study may be important only for those going into research work. Not so, say the savants! From pulpit to laboratory we are assured that science in the schools is necessary not only for future scientists, but more importantly as liberal education for the future average citizen. We physics teachers have a particular obligation to provide understanding about the basic core of all the sciences. That we do, for the 20% of high school graduates who take our formal course. Of course, even with these privileged few we cannot cover very many topics or probe very deeply into physics. As we throw out old topics from the new curricula, we assume that they are being picked up in the junior high or grade school classes where they ought to be taught. Maybe the other 80% of the students learn about physical science in those lower grades.

Maybe they do. Maybe they don't. How will anyone ever know? If a 10th grader can't read and write and do simple arithmetic, the school system knows about it and most systems do something about it. Where is the senior high school that knows whether or not an entering student has had satisfactory training in science? What standards could it use? Does this mythical high school give a diagnostic test in science to each student and then provide appropriate courses so that minimum competence in science is attained before graduation?

Some schools have established K–12 guidelines in science. However, unless testable goals are established, the guidelines are meaningless. One way to specify these goals is in terms of the minimum competency expected of the high school graduate. If agreement could be reached about that overall goal, the distribution of subgoals among the grades would be relatively simple. Particularly in the physical sciences, the sequential nature of the subject and the development of concept formation in children would determine the placement of various topics in the curriculum.

Would it not be dangerous to have such a list of overall goals? Lists of required knowledge are notoriously stultifying in education. To meet such lists, schooling often degenerates to a process of rote memorization of the required facts. Lists of noble generalizations and grand concepts sound good but are operationally meaningless. On the other hand, without some sort of commonly accepted goals, we find ourselves in our present situation. A plethora of new year-long courses has been created, each independent of what schooling went before or after. On top of that, in the grades and to a lesser extent in the junior highs, science is usually one of the minor subjects.

Who would we trust to design a list of minimum competencies in science? In the first place, there would have to be an attempt to see if agreement could ever be reached about the details of such a list. Suppose several physicists, chemists, and earth scientists, familiar with their technologies and also familiar with school realities, were to draw up separate lists of everything that they expected a high school graduate to be able to do in their field. These would be lists of *minimum* competencies. Furthermore, they would mostly consist of things the graduate could do or had personally experienced—instead of a list of facts to be known. Then let this group of scientists and educators attack each others' lists, and let each man throw out items that he does not think are absolutely essential for being able to operate in our world. The problem is, would there be any list left? If there were, we could use the list to make sure in adopting individual courses for a K–12 science curriculum that nothing important was being left out. If the list items are phrased correctly, in behavioral terms, they would compose the testing implement that would put teeth in the proclaimed goals.

Our contention is, of course, that general agreement could be found about the minimum expected competencies. We would want to make sure that the mechanism of preparing the list would provide for its annual reconsideration. If you want to see whether or not this proposal is feasible, try drawing up such a list. Say to yourself, "Here's a healthy eighteen-year-old citizen in a scientific age, not necessarily going on to college. What experiences in science are there such that I would be shocked if he had not had them? What minimum abilities are there such that without them he is incompetent to work in or understand our world?"

Here are some items from a long list that we drew up one rainy day:

1. Identify the source of energy of a flashlight, house light, car, animal, watch, tides. Make rough estimates of the energy involved for a particular period.
2. Measure and graph the position of an object versus t as it moves with constant velocity.
3. Demonstrate that the period of a pendulum is approximately independent of the amplitude, for small amplitude.
4. Measure in centigrade (Celsius) degrees, and graph, the temperature of ice slush as it is heated steadily until it has all turned to steam. From the graph, compare the amount of heat necessary to melt ice with the amount to turn water to vapor (at standard temperature and pressure).
5. Give the details of one good demonstration that matter is atomic.

6. Explain why sound can still be heard but loses its crispness when it passes around corners or between rooms.
7. Determine the focal length of a lens.
8. Choose materials and demonstrate electrostatic attraction and repulsion.
9. Measure volts and amps in simple circuits, using appropriate instruments.
10. Name representative elements in light, medium, and heavy regions of the periodic table. Describe their physical properties.
11. Be able to make elementary approximations, both mathematically and in terms of physical applications. For instance, be able to estimate lengths, masses, times, forces, areas, and volumes to within a factor of two. In simple situations be able to judge whether such an estimate or a more accurate measurement is called for.

Perhaps it would be dangerous to have only one such list with an aura of authority about it. Instead of being a floor of minimum expectations, the list might turn into a ceiling for course design. Would it not be useful, however, for all of us to draw up lists, compare them with those of other teachers, and then examine our curricula in terms of the commonly accepted items? If we find a sizable number of minimum expectations not being met by our K–12 syllabus for all students, some changes are clearly required.

November 1967

An Elementary Proposal For Elementary-School Science

many high-school and college physics teachers volunteer or are called upon to help out in the teaching of elementary-school science. We give talks in the schools, judge science fairs, loan apparatus, and run workshops for teachers. Most of us are happy to do these things because in general there is no other institutional linkage of K–12 science instruction. Neither scientists nor high-school science teachers have much input into the curriculum or texts at the elementary-school level. Can we do any good with these casual interventions? At the very least, can our visits and workshops be harmless?

The first thing to face is the hopelessness of trying to effect grand change. When the Physical Society Study Committee (PSSC) movement assayed its revolution, it needed to contact and persuade only about 4,000 high-school physics teachers. That's all there were. With federal money and professional enthusiasm, a large fraction of us got together in a variety of institutes and monthly refresher meetings. More recently, the Physics Teaching Resource Agents (PTRAs) brought over ten percent of the nation's high-school physics teachers into the revival tent. But with elementary-school teachers, the numbers are against us. There are over one million of these teachers, with a five percent annual turnover rate. That's 50,000 new teachers every year. "Great!" you say, "The new ones can learn in college how to teach science properly and so will soon replenish and invigorate the profession." There are no signs of this happening. Most people going into elementary-school teaching have a dislike for science. At best, they have studied some biology and Earth science. Even if an elementary-school teacher wants to teach science, the system is stacked against it. The commercial textbooks are filled with mistakes and have a warped emphasis on the memorization of science words. There are no standards and no required tests for elementary-school science. In any state in the union, a student can enter seventh grade and do well without ever having had prior contact with the study of science.

Nevertheless, suppose that out of devotion or the fell clutch of circumstance you enter into the Herculean labor of helping out the local school? It's your turn to run the elementary-school workshop. What should you do? First, remember that your students will teach as they have been taught. It's obvious that you shouldn't give a lecture about the virtues of hands-on learning. But there is a trap. With our high-school or college resources, or just because we're all adults, we can mount some pretty ambitious hands-on activities. Let's go visit the cyclotron, take an all-day field trip to the tidal ponds, watch the stars rise after we collect owl pellets. But it's hard to do these things with a class of fourth graders. In order to provide useful help to elementary-school teachers, we have to know the territory. Are there sinks in their classrooms? Are there flat surfaces to work on? Who will pay for the supplies? Can the children leave the room safely?

Should grade-school science be primarily nature study, or should we begin the process of building up theoretical concepts and experimental methods? For nature study you need access to nature and a knowledge-able teacher. If live butterflies can be brought into the classroom, fine! If the life story of butterflies is merely another few pages of colored pictures in the text, or a few minutes of film, forget it! No real butterflies, no real science. And there will be no real science if the teacher doesn't know how to lead the students in the observations instead of just naming the parts.

As for building theoretical models, be sure you have a realistic grasp of the developmental stages popularized by Piaget. Elementary-school children are still in the concrete operational stage. Science must be tangible. That rules out atoms, magnetic domains, solar systems, genes, and energy conservation. Of course, it rules out Newton's laws. "But," you say, "my own sixth grader knows that energy is conserved." Of course, she does. So did mine. All our children are precocious. They can all answer correctly that energy is conserved. But, like most sixth graders and some twelfth graders, they don't yet know what energy is.

What's left? If we can't bring in butterflies and if the children and their teachers can't solve simultaneous equations, how can physics teachers help in elementary school? We can teach them to measure. The symbols of our profession are the meter stick, the balance, and the clock. From kindergarten on, all of natural science can be studied quantitatively, using the math appropriate for that grade level. Note! Science now becomes entwined with math. Elementary-school teachers have to teach math anyway. The science link makes the math tangible, and thus makes teaching math easier, and thus makes science palatable.

Physics teachers can devise some very realistic and useful standards for science accomplishment in the grades. We merely ask what we want our students to be able to do when they enter our classes. Wouldn't it be lovely if they could properly use the standard measuring devices? Wouldn't it be inspiring if they had graphed so many measured variables over the years that they could recognize graphs of the simple functions? Wouldn't it be exciting if they had handled so many phenomena that they knew the approximate sizes of many things? Would it be asking too much that they know these sizes in SI units?

Students can learn before they are twelve years old how to use the standard measuring tools. There are no conceptual barriers. They can learn to graph relationships that are interesting and useful to them. They can literally handle in a quantitative way a large number of fascinating phenomena in all the natural sciences. Their teachers can learn how to do these things too. The required apparatus is inexpensive and easy to use in standard classrooms. Children like to use these tools, partly because in most cases they have to get out of their seats in order to handle the phenomena and make the measurements. It is feasible to establish standards and performance tests for these skills, all of which are fundamental to future science studies.

Perhaps elementary-school science is in bad shape because we have been trying to do too much. Some programs try to make little scientists out of the kids. Let's just teach them some simple measuring skills. There will be time for them to learn quantum mechanics and the names of the planets in seventh grade.

April 1990

L ast April on this page I opined about appropriate science instruction in our elementary schools. Raising my sights, it is only fair
that I now pontificate about the junior highs. I do not propose to
list what seventh, or eighth, or ninth graders should *know* about science. I am, however, concerned with what they are able to *do.*

To start my analysis scientifically, I made diligent inquiry into the
science instruction of two young people who happened to be guests in
my house. Of course, these two children are particularly precocious, as
I'm sure your grandchildren are too. The fourth grader was learning to
classify things as either plant or animal, and had just got through learning that scientists believe that there are atoms although you can't see
them. In the April editorial I denounced such wanton silliness, and so
shall weep here no more. The seventh grader had finished the experiment (done by the teacher) of growing bean plants surrounded by red,
or blue, or green cellophane. The class thought that the plant inside the
red filter had done better, but of course had made no measurement of
the light intensity getting through the material. (I thought that criticism
of that stupid exercise had killed it off thirty years ago, but apparently
the word hasn't gotten around yet.) At home these two children learn
that if it isn't quantitative, it isn't science, but they know that school is
strange and different. Fortunately, the seventh-grade class finished biology and has now moved on to the six forms of energy. My grandson
could remember only five, which was one more than I got right. *Wind
energy?*

So then. I agree with Bill Aldridge and his *Scope and Sequence* plan to
include some physics at every grade level. Naturally, the milieu must be
such that the children have fun and learn the scientific method and
don't wreck the apparatus. In each grade I would have some theme to
tie units together. In seventh or eighth grade the theme might well be
energy transformations, though the goal would not be the memorization of the six forms. I would start by having students perform (with
their hands and bodies) various chores requiring physical work, such as
lifting heavy boxes from floor to table. This work should be quantified

in terms of a measurable unit, and that in terms of other measurable units, all SI. The children should then learn how to do this work with large-scale simple machines, measuring the appropriate forces and distances. Work could then be stored in the energy of raised weights and compressed springs and then partially recovered with efficiencies measured and calculated. Energy would become a tangible thing that you must put into a storage device (perhaps a rubber band) with the work of your hands and which then turns into other kinds of energy. The theme might lead students to make materials hotter and so practice some of the experimental techniques of measuring temperature. It might turn to the subject of batteries and electricity, where students would learn how to handle the first and second laws of electricity. (1. All wires have two ends. 2. They may be insulated.) In the Intermediate Science Curriculum Study (ISCS) program, seventh graders made their own small storage batteries and then used them, while the energy lasted, to power motors and light bulbs. When the energy was gone, they had to charge up the battery again. The energy theme could bring in useful skills of daytime astronomy. For instance, the study of solar energy requires measuring the power density in W/m^2 and doing this in terms of the changing orientation of the Sun. Human biology can be brought into the theme by measuring heat output and energy intake of students.

Many elements of such a program have been done before. All that is needed is to make sure that the students learn to do things besides sitting and watching. In order to bring that about, we need to define our goals with tests. The tests must necessarily be hands-on performance tests to see whether the students can do things. Here are some of the things ninth graders ought to be able to do.

- Measure lengths with meter sticks and by pacing, and know when to use which.
- Measure masses with various balances.
- Measure time with ordinary clocks and with short-interval timers.
- Measure temperature with various types of thermometers, and be able to insulate objects or conduct heat rapidly between objects.
- Graph the dependence of one variable on another, including making and labeling the graph and its axes with appropriate units.

- Make one-step calculations using power-of-ten notation, and use a hand calculator to check the results.
- Use wire, battery, and bulb with the right tools and connectors to make the bulb light.
- Set up a convex lens as a magnifier, and produce both real and virtual images.
- Measure the volume and mass of an object and calculate its density.
- Choose the right instruments and measure the speed of a student running in a race.
- Cover the mouth tightly with a tissue while coughing or sneezing.

The list goes on and should encompass all fields of science. I suppose that it's possible to test some of these skills with multiple-choice questions, but I doubt that it would ever be done well. Note the great absence of scientific facts to be memorized. Note also that I do not require the student to display his or her attitude toward the subject or its practitioners, nor am I concerned with how that attitude has changed as a result of the course. Since we already have the popular TOUS (test on understanding science), perhaps we could call the new test, TODS (test on doing science), or GOOYSADS (get out of your seat and do something).

The main difference between the requirements for junior high and for elementary school is that I assume that in junior high there is specific time scheduled for science in science rooms with science teachers and with some science apparatus. Furthermore, I assume that the students are now old enough to have been baptized, or confirmed, or bar mitzvahed, or have otherwise reached the age of reason, and thus can slowly be introduced to simple analogies and functional relationships. No atoms yet. There's nothing junior-high students can do that requires the concept of atoms. Save something for high school.

If you don't like my list, draw up your own. Just write down all the things you would like students to be able to do (not know) when they enter your physics class. They might as well be taught these skills in junior high; puberty is a terrible thing to waste.

March 1991

More Sage Advice—The High-School Physics Course

nyone who knows what should be taught in elementary-school science (April 1990 editorial) and junior-high physical science (March 1991 editorial) is clearly a recognized expert in what should be taught in high-school physics. When truth and wisdom finally prevail, remember that you heard it here first.

Notice that fewer than one-fifth of our high-school graduates study the formal course in physics. Second, note that the standard course looks very much like the standard introductory college course. We all teach Newton's laws of motion in the fall and Ohm's law in late winter, followed closely by some nodding introduction to atoms, transistors, and these days, quarks.

People have been criticizing this situation for years, and very recently groups have been moving to change it. At the college level the Introductory University Physics Project (IUPP) has been sponsoring and encouraging the development of radically different approaches to the introductory course. One of these proposals jumps into twentieth-century physics with an emphasis on solid state. Another concentrates on nuclear and particle physics. At the State University of New York at Buffalo, Jonathan Reichert has been teaching a first-year course for engineers and physicists that starts right out with the relativistic equations for particle dynamics. It would be a prudent idea for students in such courses to have had a preliminary high-school course of the traditional kind. However, they may not get it. There's a parallel movement that started with summer workshops at Fermilab to introduce modern (particle) physics into the high-school course. We had a session (AB) on that subject at the annual meeting this past January.

Meanwhile, people outside AAPT have been busily moving into our territory to remedy our lack of attention to the other four-fifths of the high-school population. Those of us with long memories may recall that *Harvard Project Physics* was supposed to appeal to some of these students—the ones who didn't like *PSSC*. It turned out that for either course you had to be able to read, and so both were drawing on the same population. Then there was *Man Made World*, sponsored by the engineers, which contained marvelous curricular material, including instruction on feedback, but was aimed at the tenth-grade level in which there was no slot and for which there were no teachers.

The new programs are supposed to be designed to fit into the standard curriculum and be attractive to students who wouldn't take physics. One of these is *Technology Education*. We had a brief description by John Roper of this program in the January 1989 issue of *TPT*. Its originators and sponsors come from the ranks of those who used to be called shop teachers, but who are now teachers of industrial arts. The course is aimed at students who do not take our formal physics course, either because of the math demands or because of a perception that our subject has no immediate practical use.

An even more ambitious, though currently nebulous, plan is National Science Teachers Association *Scope and Sequence*. Millions of dollars have been committed by NSF and the Office of Education to back curriculum development that will weave physics into each grade, 7–12. (An old school motto was "Every class an English class." The new motto will be "Every science teacher a physics teacher—somehow.")

There is yet another program aimed at students who do not take our standard course. To some extent it is modeled after the successful new high-school chemistry course called *ChemComm*. The subject matter is supposed to be of immediate student concern, or at least *community* concern—environment, and all that. Similarly, the "Just Physics" program, if National Science Foundation funds it, will teach physics of the everyday world of students. (The first thing to do, if it gets funded, is to run a contest for a better name.) American Association of Physics Teachers and American Institute of Physics jointly mounted the proposal.

For the standard course there seems to be two competing philosophies. One derives from the cautionary tales provided by learning theorists. Many high-school seniors have not yet fully arrived at the formal analytical stage of reasoning that we assume when teaching the standard course. No matter how cleverly students learn to manipulate $x = \frac{1}{2}at^2$, many of them still think that at the top of a trajectory a ball's acceleration is zero. The opposing view is that since all the other high-school students in the world are apparently ahead of ours in solving test questions we should make our physics and math courses more rigorous and time consuming.

For the last three years I've been a consumer of the high-school product; I've been teaching an introductory university course to students who took high-school physics and did well in it. Here's what I think we ought to aim for at the high-school level. 1) Assume that the students will take no more physics; this is our last chance to teach them the nature of the physical universe. 2) Do not try to cover the canonical subject and do not worry about the standard sequence of topics. 3) Concentrate on teaching quantitative methods of solving interesting problems. The ability to deal with the difference between parts per million and parts per thousand is worth more than knowing Newton's laws. 4) Provide hands-on experience with the sizes of many things, expressed in standard units and manipulated with order-of-magnitude calculations. 5) Teach the use of the principle of feedback in various phenomena. 6) Use analysis that requires geometry, trig, and algebra, and talk the math department out of offering calculus, since clearly most high-school seniors aren't old enough for calculus. 7) Make sure that seniors know the scale of their universe, from atoms to the edge, and from T_0 to now, and have become operationally familiar with appropriate mileposts along the way. 8) As a paradigm of the sort of topic that should be included, let no student graduate without being able to use Archimedes' law to explain how a ship floats.

The trouble is, Archimedes' law is not part of the standard curriculum. If we put something in we must take something out. You want my advice? Take out quarks.

April 1991

What Do College Professors Want?

A perplexed Freud once asked, "What do women want?" Teachers at all levels may well wonder what society expects of them and their students. Should the student be well-rounded culturally, germ-free and agile physically, deeply versed intellectually, active (but not activist!) politically, and also happily content psychologically? And as far as we are concerned, should the student also have a working acquaintance with all of the standard topics in mechanics, heat, light, electricity, atomic physics, and the history and applications thereof? All of this activity should not interfere, of course, with senior play, band rehearsals, and team athletics.

On top of society's demands, teachers of sequential courses face special expectations. Many, if not most, high-school physics students will take a college course in physics. The role of the high school course is not just to be a lead-in to the college course. The high-school experience should be complete in itself. However, if many high-school physics students continue the study of physics, everyone along the line should be interested in the continuity of experiences. What does a college physics professor expect of students who have already had a high-school course? Eleven years ago a group of distinguished college teachers gave their answers in an article in *The Physics Teacher* (4, 218 [1966]). Here are some thoughts gleaned from recent conversations with teachers of the introductory college course.

Some professors claim that no prior physics work is required for their freshman course. Everything is presented sequentially from first principles. Of course, the same assurance could be given for a graduate course in quantum mechanics. My own experience has been that students who did not take high school physics are at a serious disadvantage in any of our introductory college courses. Regardless of their memory—or lack of it—concerning particular details, the students always seem to understand the logic, the methods, and the language better the second time around.

Certain skills are particularly important in college physics. We might be tempted to speculate that the skills should be taught directly instead of as a by-product of the high-school course. No one ever seems to do it that way; perhaps it would be hard to make such a course interesting. Furthermore, an earlier exposure to the standard physics *content* seems to help in subsequent studies. What helps even more, if the student is lucky, is an introductory course that arouses curiosity and healthy skepticism. Every teacher would like to have students enter his class with burning questions.

Let me list the skills that we would like to see in our freshman students, and also point out what they don't need. First of all, in laboratory they should have lots of experience in handling measuring instruments but they have little need for writing formal lab reports. They should know how to avoid parallax in readings and have some appreciation for the significance of significant figures. No one expects youngsters to use good judgment about the degree of precision that is appropriate in a particular measurement, but it would be useful if freshmen were at least aware that such a problem exists. It would also help if all high school graduates had a feeling for screwdriver directions, and knew the proper way to fasten wires to binding posts. A good shop course would help.

In fact, it would help if physics students had a rich store of first hand experience with phenomena. They should not only have seen rainbows, but observed them, and know where to look and what colors to expect. Knowing about wave-particle duality is not so important for our purposes as having watched ripples in ponds and being familiar with camera, or telescope, or eyeglass optics. While we might hope that everyone nowadays knows that electric current is in the direction of positive charge flow, all theoretical knowledge of lines of force can be forgotten if the students have retained the ability to wire circuits and use voltmeters and ammeters.

We would trade in all definitions of terms and memorized laws for experience in using the simplest of trig and algebra. How happy physics professors would be if all their students could sketch the graphs for the sinusoidal, power, and exponential functions, and knew how to express correctly the arguments of those functions. For instance, no math course at any level points out that the derivative of $\sin \theta$ is equal to $\cos \theta$ only if θ is measured in radians. Furthermore, it's in physics, not math, that students learn how to find and combine vector components, and to deal with the geometries of spheres, cylinders, and cubes. To learn how to use math, real data taken by the student should be analyzed, manipulated, and graphed.

As far as freshman physics is concerned, forget atoms, quantum mechanics, and relativity in high school. What our students need is a muscular acquaintance with the metric units, and free-wheeling experiences in estimating sizes and doing order-of-magnitude calculations in their heads. We hate to see youngsters cry when the batteries die in their calculators.

Wouldn't it be wonderful if every college freshman came equipped with these basic skills? I showed this list to a colleague who's teaching one of our first-year graduate courses. He agreed. "If only," he sighed, "every graduate physics major possessed such basic skills!"

January 1978

Impedance Matching

When a signal crosses a boundary between two media, part of the energy is transmitted and part is reflected. To minimize reflection, you should try to make the impedances of the two media equal or create an impedance-matching boundary layer. Thus we have flaring bells at the ends of musical horns and purple coatings on camera lenses. To be sure, if there are no reflections from the ends of the horns, we get no music, but usually we seek smooth transitions.

I've been visiting high-school physics classes recently, not to teach but just to sit back and enjoy the different milieu. It's a rare privilege to be able to do that, to observe experienced teachers in action and to watch and listen from the students' side of the room. I have seen high-school teachers preside skillfully over blackboards and demonstrations in the traditional fashion and sat in on student group activities conducted in a less formal mode. Regardless of the format, I have been struck by the fact that the milieu, the environment, the culture are very different from that in the university. Of course, the students are slightly younger than the ones I teach, but in less than a year most of these high-school physics students will have passed the boundary into college. How smooth a transition will that be for them? How well do we physics teachers match impedances of the two levels?

All over the country, college physics teachers are concerned about the mismatch between the requirements of the standard first-year course and the abilities of the students. We have been holding conferences and each other's hands, wondering if we are requiring too much or if there really has been a change in the students. In large university settings, it is not unusual for 40 percent of the students in the introductory physics course to fail or drop out. Some universities have created special transition courses that are prerequisite to the standard physics course. Others have stretched two semesters of work into three. A number of groups in American Association of Physics Teachers are designing new types of introductory courses, all of them to be shorter, if not easier, than the traditional course.

There are some things that might be done on the high-school side of the boundary to ease the transition. Apparently, most high-school students don't read their physics texts. Some teachers complain that the texts aren't good enough, others tell me that students just won't read, and some claim that their students can understand physics better through labs or films or other activities. College courses, however, rely heavily on texts. The professors assume that the assigned reading will be read. They furthermore assume that the students know how to read. There's a boundary layer! It takes special skill and experience to read a college technical text. The chapter should first be skimmed to get the general intent of the author, with special attention paid to the introduction and conclusion. Then it's necessary to read the material with pencil in hand, reproducing the derivations on a separate sheet of paper. Underlining or highlighting is worse than useless in a physics text. It is appropriate, however, to note in the borders the points you don't understand, which you will pursue later with classmates or teacher's assistants or professor. There should be specific lessons in high school to teach the skills of reading technical texts. This applies particularly to bright students who have never had to bother with texts.

No matter what the high-school teaching and learning format, and no matter what the current learning theory fad, in college most students are going to sit in large lecture sessions. But most freshmen don't know how to take notes and use the lecture format. The rough notes taken during the lecture should serve only as a first draft for the revised notes to be written later. During the rewrite (to be done within the following 24 hours) the student should mark places where things don't make sense or which need further explanation. Those points should then be raised with classmates or instructors. For successful impedance matching, high-school physics students should receive specific instruction and practice in taking lecture notes. The bright students particularly need this drill.

There are some minor adjustments in the high-school program that might help the transition to college. For instance, the word "error" is a technical term meaning the amount of uncertainty in a measurement. Error is not the difference between the student's measured value and the "accepted value." Indeed, the very term, "accepted value," should be anathema to physicists. Accepted? By whom? Who measured it? What was *their* error? Consequently, the term "human error" is misleading, if not meaningless. It should never be used. Probably, "random error" should not be used either, unless you plot the data and observe a Gaussian distribution. The difference between my measurement and a handbook value is called a discrepancy. If my measured value with its error does not overlap the handbook value with its error, I may worry about the discrepancy. It takes a lot of training and experience to judge the appropriate magnitude of an error. High-school students may not be old enough to worry about the fine details, but at least they should get the language straight.

Another small point is that "mechanical energy" is a rather precious term meaning precious little. It's the sum of kinetic and potential energy, but to dignify it with that particular name leads to wrong impressions. Potential energy, incidentally, is not a synonym for gravitational potential energy. As for Einstein's famous equation, it does not describe mass turning into energy; mass and energy are the same thing. And while we're being really petty, it might help the cause if at least half the time students would refer to "g" as the gravitational field strength instead of "the acceleration due to gravity."

There may even be some petty points, or rough edges, concerning the other side of the boundary. I have heard rumors of college classes taught by instructors who are slightly unfamiliar with the local inhabitants perhaps even with the local language. I have been told of college physics courses that concentrate on formulas and problem solving to the exclusion of the romance and mystery and excitement of the subject. Naturally, I find these allegations hard to believe, but impedances do indeed depend on an inertia term.

Whatever the reasons, many students have trouble with the transition between high-school and college physics. The teachers on both sides of the boundary should actively seek information about what the world is like on the other side. The more we know about the problems and requirements and practices of our different environments, the more we can prepare our students to cross the boundary and the better we can mold the boundary to make the transition easier. Surely physicists, of all people, can learn how to match impedances.

March 1990

Now then, students, watch carefully the symbols in my left hand while with my right hand I draw from this derivation the most breathtaking conclusions, inferences, and generalizations. Yes, Jones? You missed the point there, eh? I had rather thought that it was intuitively obvious. Why not think about it some more? Actually, the full proof can easily be shown, but is beyond the scope of this course. If there are no further questions, then, class dismissed.

Fortunately for our reputations, most students will swallow anything, and unfortunately, so will we. It's a common saying that the best way to learn something is to teach it, and the best way to learn to teach it is to write a text about it. In the process we discover, time and again, that the explanations that have seemed so satisfying when we learned them are filled with gaping holes. Graduate students, in their first year of teaching, begin to understand freshman physics for the first time. (Usually they are so fascinated that they complain, "Why weren't we taught these things?" Of course, they were.) The embarrassing truth is that there seems to be no end to the business of learning that we don't know things. Even after 35 years of teaching I find that I don't really understand some of the most common phenomena that I have been glibly describing all this time.

We don't usually intend to fool our students with logical sleight of hand. In many cases we don't even realize that our explanations are flawed. Our left hand has fooled our right hand for so long that everything seems perfectly reasonable. In other cases we may get an uneasy feeling in the middle of a lecture that we don't quite believe what we just said. The moment passes. We are nimble and the students are laggard. We make a mental note to worry over the argument some other time, preferably before next year. Frequently if we do pursue the matter, we discover that many of our colleagues also have trouble with that particular explanation. Perhaps no one has an explanation suitable for an introductory level. Let me illustrate.

What could be more fundamental than understanding how electromagnetic fields are emitted by a charged particle? After all, that's the topic of our second semester. How do you explain the field of a moving charge some distance away? When do the field lines point toward what? If the charge is accelerated, how is the information propagated outward? Oh, there are diagrams to use of kinks in radial field lines, but have you really followed through the details? No need to figure it out this year. A kink in time saves nine questions, if you wave your hand and hurry on before the students notice.

A common feat of scholastic legerdemain is to characterize some unknown phenomenon in terms of another one that is supposed to be well known. This trick is called illusion by allusion. Thus electrical circuitry can be taught by referring to the well known behavior of water in pipes. The method is canonized at the next level of instruction by using the terms *divergence* and *curl.* Unfortunately, our students are as unfamiliar with the flow of liquids as they are with the flow of charges. Furthermore, we ourselves may not really be familiar with familiar examples. How many of us (and how many texts) refer casually to the mechanics of billiard ball collisions? Have you really studied the behavior of billiard balls? I, for one, have not, but I have discovered that such collisions are dominated by spin and the reactions of the felt covering on the table. Certainly the complex behavior of billiard balls makes them a poor model for the simple collisions we study in introductory physics. But if we mention the well-known billiard balls and rapidly move on, we have let students know that physics is about familiar things, which they don't happen to understand, although apparently everyone else does.

We all know how simple and useful Huygens' principle is for explaining wave motion. Why don't the Huygens' oscillators propagate waves in the backward direction too? Very good question. Are there any other questions? (If a student really wants to know, refer her to "Questions Students Ask" in the February 1980 issue of *The Physics Teacher.*)

Everyone knows how high and low pressure regions travel across the country. But what happens to a low pressure region over the mountains where the normal pressure is already low? (In Denver, at 1.5 km above sea level, the normal atmospheric pressure is 63 cm, or 24.8 in., of mercury. Not even the eyes of hurricanes get that low on the coast.) If children worry enough about memorizing the weather map symbols, they won't have time to ask such questions.

Then there's the all-time show stopper of swinging a pail of water overhead. Centrifugal force keeps the water from falling out, right? The calculations prove it. So this kid in the front row says, "Maybe at that speed there isn't time for the water to fall out." Well, of course he's right, but why confuse a perfectly good argument? Duck, feint, and move on to the next illusion.

One of my favorite lectures allows me to demonstrate my aging skill with a bugle. With oscilloscope, signal generator, string, and meter stick I relate the allowed, quantized frequencies of the bugle to its length and the speed of sound in air. True, I cannot sound the fundamental (the pedal tone), which would rather spoil the argument if I dwelt on the problem. But do I belabor this slightly embarrassing fact? Nonsense! The explanation is technical and beyond our present scope. Onward to standing waves, and vibrating lips triggered by reflected pulses. After the lecture, I sometimes amuse myself by playing recognizable bugle calls with just the mouthpiece. And I briefly wonder each year how that's possible.

At least we're on solid ground with Newton's laws. No tricks there. Clearly F is proportional to a and m is the proportionality constant. What's F? Why, it can be measured by accelerating a known mass and thus you can calibrate a spring. How do you know the mass? Simply exert a known force on the unknown mass and measure the acceleration. Is there no way out of the circular argument? Of course, there are many ways out, each tricky, and most not really valid. Whatever you do, don't confuse the student. He has enough trouble trying to remember whether m is F/a or a/F.

In our September 1970 issue, Mark Zemansky taught us the subtleties of using the word *heat*. He reminded us that:

> *Teaching thermal physics is as easy as a song.*
> *You think you make it simpler*
> *When you make it slightly wrong!*

After all, isn't our purpose not so much to teach the truth as to leave a truthful impression? With almost all of our tricks of explanation, there is a deeper, more complex level. Below that level, there may be another.

The more fundamental the concept, the more the complexity. I would explain all this to you more fully, but it is beyond the scope of this editorial, and since the fact is well known and can easily be shown, I will leave it as an exercise for the reader.

January 1980

The choice I faced was between writing this editorial or spending an afternoon on the ski slope here at West Point. It was a clear choice between duty and pleasure. A short time later, as I rode on the T-bar to the top of the mountain, I promised myself that I would nevertheless think about physics and physics teaching all the way down. Why, for instance, can heavier people go faster than lighter people down a ski run, and why can you go faster with longer skis? What is the angle of an incline that looks from the top like a precipice, and what is the insolation in February on a northern-facing slope? What's the shape of snow flakes coming out of the artificial snow machine? These are all good thoughts for a physics teacher on a downhill run.

Since speed actually does depend on weight, I conclude that some of the friction forces must not be proportional to mass. That's certainly true of air friction, which is important. In spite of our textbook examples, the surface friction of skis on snow depends on the surface area of the skis. Longer skis are faster. Evidently a large surface area cuts into the snow less. As for the steepness of the hill, a 15° or 1/4 radian, slope means a one meter drop out of four. That would produce an acceleration of $2\frac{1}{2}$ m/s². In ten seconds, straightaway, you would be going 25 m/s, or about 55 mph. Some of the trails are certainly steeper than that, but not the ones I go on. In February, at our latitude of 42°, the sun rises to 40° above the horizon. On a northern slope of 15°, the angle between the sun's rays and the normal to the surface is 65°. The cosine of 65° is 0.4. Hence our snow should last awhile longer if it doesn't rain and if the temperature at night goes below -2°C for the snow machines to work. I don't know what artificial flakes look like. You want to know how the snow machine works? Pretty well, I would say.

A cadet rides up the T-bar with me and learns that I teach physics. He's in another section. "Doing all right," he says, "but there's an awful lot of formulas to learn." He asks me why we define magnetic field lines to be perpendicular to the velocity of the electric test charge. Why not just plot the forces on a charge at various points and say that those are the field line directions? It's a good question for a T-bar and I try to answer at the top with a diagram in the snow. He's polite, all cadets are polite, and thanks me as we wish each other a good run. But all the way down I wonder if maybe we should start the subject of magnetic fields with permanent magnets. That's the way we used to do it. Then we plotted the fields with a north pole isolated at one end of a magnetized knitting needle. Some lessons later, after the field patterns from pole to pole were established and familiar, we learned what happens when we put currents in those fields. The logic was not so clear as in our present system, but maybe the pedagogy was better.

The next time up I ride with another cadet, one of my own students. I ask him how he likes studying electricity. It turns out that the subject isn't what he expected—Gauss' law and all that. He wanted to learn how to wire things. In high school he had learned about charges and the right-hand rule and even about cyclotrons. But he had never really wired anything. It was like last fall in mechanics where he had learned about cross products and torques. But he had never rigged a pulley. I tell him to forget it. He's going to be an officer. He can tell other people to wire the circuits and rig the pulleys. Up at the top we wish each other a good run, but I feel guilty all the way down.

April 1982

The Bottom Lines

This past month I have been on a panel to judge samples of science writing for an annual contest sponsored by the American Institute of Physics. Most of the contestants were professional science writers. The entries consisted of books for a general audience or articles in major newspapers or magazines. In most cases the writing was very good, well aimed at its particular audience, and scientifically accurate. It was amusing to see how the writers and editors viewed their readers. For some magazines there had to be a strong grabber in the first line with florid adjectives, and an anecdote (I peered through the giant lens . . . .) every other paragraph. For more pretentious journals the style became more formal, even to the point of using references, if not footnotes.

What sort of criteria are appropriate in judging science explanations for such diverse media? Readability, for one. The writing must be lively enough to capture and keep the reader's interest. Correct level and correct science, for another. In order to reach the right level it should never be necessary to make the science slightly wrong. (When that happens, it means the author doesn't understand the subject well enough.) It seems to me, however, that there is a bottom line to be considered. Will the reader learn and remember something? If not, why not just have another page of comics, or why not just write another novel?

Now, the bottom line for most publishers is not the educational efficacy of the product. First and last, does the product sell? Even the most altruistic publisher must make sure that the books or magazines or newspapers are bought. For that purpose, most science writing need merely give the reader a good feeling. There won't be any test at the end of the chapter. But the reader, or a critic, should demand more. If the writing is really good, then it will be instructive. The reader will learn something and perhaps be able to do something or think something new.

While pondering those noble thoughts from my judicial eminence, it occurred to me that I might well use the same criteria to judge myself. What's the bottom line for evaluating my own teaching? Well, of course,

I had better make sure that I continually keep in touch with my audience, to make sure that I'm not kidding myself about their level of understanding. Then I should continually review my explanations and analogies to make sure that they are valid and that I'm not teaching science fiction. For all sorts of good reasons I should pay attention to the bottom line of class registration. If students don't sign up for my course, my lucid brilliance will go unappreciated. Worse yet, it may go unrewarded—administrators have their bottom lines, too. Every carnie knows that you can't sell candy if you don't first get the rubes into the tent.

But we teachers have another bottom line, complete with tests. Does our audience learn? We ought to be concerned also with a more difficult question. Will our students remember beyond the test? Has the course changed them? Has it made them more skillful, more powerful intellectually?

Teachers have an enormous advantage, and terrible responsibility, compared with authors. Besides the facts, our students are studying us. After they forget Newton's laws they may remember our attitude toward learning, our style in facing problems. Every so often throughout the year, perhaps particularly now in the rushed days of spring, while there is still time before the final exams, we ought to review what's happening to us and to our students. Consider the bottom line.

April 1985

Homely Physics

We have introductory physics, fundamental physics, conceptual physics—why not homely physics? Actually, we have a lot of homely physics in *The Physics Teacher*. Ron Edge keeps showing us ways to demonstrate physics with only string and sticky tape. Earl Zwicker and the Chicago gang remind us about marvels and paradoxes to be found in toys and simple devices. Dick Crane reveals the physics inside the gadgetry of everyday life. Of course we also carry articles about exotic research concerning particles and galaxies but it's useful to test our skills and sharpen our wits on more mundane mysteries.

When we revised the New York State high school physics syllabus 20 years ago, we threw out, among other things, Archimedes' principle and simple machines. Elementary material like that should be taught in junior high. However, the junior high physical science teachers are too busy teaching about nuclear physics, Coriolis forces, and atomic orbitals. The kids never do learn about floating boats and pulleys and levers.

In our introductory university course this year I have attempted to remedy this situation with special laboratory assignments. We still have many of the typical lab exercises, using air tracks and other standard school devices. But each week the students must also do a homely measurement. It is homely in several ways. First, the measurement must be done at home or in the dorm. Second, the phenomenon to be investigated is common, everyday, though not always understood or appreciated. Finally, the measuring tools must also be common and cheap.

Student reaction to the homely exercises is mixed and revealing. At the beginning they felt a little foolish dropping balls from windows or pacing the halls. By the middle of the term, however, most students were making games of the problems and were involving family or roommates. Physics, crude but real physics, was being done all over campus. Still some students felt that what they were being asked to do was not real science. After all, the apparatus was homemade and high accuracy could not be obtained. For instance, some students gave up on one exercise which required them to measure *g* by timing the drop of a ball. They claimed that they didn't have access to stop watches. They really weren't convinced that short intervals can be timed to a fifth of a second by saying "One thousand and one, one thousand and two. . . ."

The crucial feature of these homely exercises is that each requires measurement, not just the qualitative observation of phenomena. When you are using crude or homemade measuring instruments, the nature of error—uncertainty—takes on a homely reality. For example, the assignment to measure g followed a laboratory session where, with special apparatus, g could be measured to $\pm 2\%$. In the home exercise the students were required to measure g to $\pm 30\%$ by timing the fall of an object. Furthermore, the entire actual measurement had to be done in less than 15 minutes. The combination of time and precision constraints makes the problem nontrivial. A feasibility study and planning must be done in advance.

Other homely lab assignments have required the measurement of each student's ball-throwing speed (without measuring θ_0 or time); the floor area of the physics building, or dorm (in no more than 15 minutes for the actual measurement); the period of a simple pendulum as a function of amplitude; the speed of water shooting horizontally from a hole in a container, and the relationship of that speed to the water depth in the container; the density of the student (measured in tub or pool); and the free period of the student's arms and legs, and relationship to the periods while walking.

We also require students to observe and describe phenomena qualitatively at home or in the dorms. These qualitative observations are intended to expand upon lecture demonstrations. We want to emphasize that physics can enhance the appreciation and understanding of the everyday world, and that such observations do not always require complicated school apparatus. However, the purpose of the homely *lab* exercises is slightly different. We are not so much concerned with learning about phenomena as we are with learning how to do things quantitatively. We want students to learn how to tackle a measurement problem as a series of approximations; first, order-of-magnitude, then one significant figure, etc. At each stage they should ask what precision is needed and how expensive, in time or money, the next stage will be. We want students to think of measurements in terms of "a 30% measurement" or a "5% measurement." They should get the feel for the difficulty of doing precision work, and should gain enough experience to know that you shouldn't try to get more precision than you need. That sort of realization is more important than knowing which is Newton's second law, and will probably stay with the students a lot longer.

Our profession deals with exalted themes. Our telescopes look back ten billion years, our accelerators probe the nucleons. Every first year physics student should get occasional glimpses of this grand enterprise. In the meantime, most of them still have to learn how to hold a meter stick in order to eliminate parallax. The way to prepare for the stars is to do a little homely physics.

January 1985

Promises, Promises . . .

t is a popular myth, much fostered by Democrats, that in 1928 Hoover promised the voters a chicken in every pot. (Actually, it was Henry of Navarre, in 1600.) In 1996 Clinton promised that by the year 2000 every classroom in the nation will have computer access to the Net. Desiring to maintain an absolutely neutral position, I think I can honestly say that I am in favor of both promises. However, as far as I can see, neither one has anything to do with teaching or learning physics.

As for the chicken, of course you can't learn physics on an empty stomach. As for computers, of course students should learn touch typing and the location of the esc key. Modern technology in schools, in industry, and in everyday life requires computer familiarity. With friendlier programs it's easier than ever to attain proficiency these days, and many of our students are better at it than we are. There are tutorial programs for solving problems, and programs for creating individualized tests. There are programs for data collection, data analysis, and graph plotting. There are programs for simulating physical phenomena, including the creation of interacting systems. There are CD-ROMs that contain enormous libraries of physics information, both in words and in pictures. And now we can surf the Net, bringing in up-to-date information from all over the world. Students and teachers have never had it so good.

We get manuscripts from teachers who have experimented with all sorts of new methods and curricular materials. It's time we had instruction and testimony from teachers who have computerized their physics classrooms. How does it actually work in practice? How much did it cost to begin with and what are the maintenance costs and problems? How much time is spent with malfunctioning machines and bewildered students? How many computers do you need for how many students, and how many connections to the Net are needed? Do the simulations take the place of old-fashioned lab materials, and do the tutorial programs take the place of other homework? Are there computer programs that are effective in the development of the type of concepts tested in the Physics Concepts Inventory? What information do students need from the Net or the CD-ROM that they can't get from their textbook or library? Is the teacher still necessary or would a skilled hacker do as well?

All of us would like to hear from any of you who have experience in these matters. It would also be nice to know the bottom line. How much does it cost? How much more capable are the students after they use it? How do you know?

Do you suppose that by surfing the Net, a student could figure out why that chicken in the pot is temporarily quite stable at 300 K, but at 373 K rapidly turns into something lip-smacking good?

November 1996

Fatherly Advice

.K., here's your assignment. Give an introductory talk to a couple of dozen new teaching assistants (TAs) about how to teach labs. That's what I had to do toward the end of August. We bring our new graduate students in early for a week of orientation about their dual roles as students and as teachers. What would you tell them about teaching? Here are my notes. Maybe I've left out what you think should be most important. Maybe I have emphasized what you think is unimportant—or unattainable. Let me know.

First, I told the TAs, you've got to know your students. That starts with the trivial, but vital, business of learning their names. If you're gifted in doing that, fine. If not, take pictures of them and study the photos. Then use their names when you talk with them.

And talk *with* them, not *to* them. Individually. And listen. If you are asked a question, don't be too ready with *the* answer. Find out where they're coming from. Temper the wind to the shorn lab. Of course, if they just want to know where to plug in the cord, there's no need to engage in a Socratic dialogue.

Every teacher should be aware of the general findings about conceptual development that have been publicized under Piaget's name. (Joseph Henry, of inductance fame, warned us about these same conditions 150 years ago.) You can't teach a six-month-old baby to walk, regardless of teaching methods—lectures, group discussions, authentic assessments, beatings. Research (or listening to students) has shown that there are similar developmental problems all the way into college. There's a good chance that some of your freshmen, even in the calculus-based course, are not very sure of algebra. You won't find this out on multiple-choice tests. Listen. Observe. Individually.

With my group of TAs, I asked everyone, and thus no one in particular, how they would explain to a freshman about the difference between inertial and gravitational mass. By the end of three seconds, no one had raised a hand. Next I suggested that they form groups of four and take two minutes to work out an answer. Of course, then they did pretty well, though most of the groups wanted to demonstrate that they knew the difference, not how to explain the difference to students. We then had a little discussion (not lecture from me) about the two different ways I had asked the question, and the moral of the tale.

Every physics talk should have a demonstration. I have a papier-mâché model of a human head, with a removable lid. I used this to simulate how useful a lecture can be if you want to fill a student's head with knowledge. With many flourishes, I poured water into the top of the head, but it all leaked out.

These graduate students teach our introductory labs under the supervision of the faculty member who is the recitation instructor. Regardless of how good or poor the recitation is, there are some particular problems connected with lab teaching. The sine qua non requirement is that the TA must have done the experiment first, using the same apparatus that the students will use. From this experience, the TA may have some tips or instructions to give to the class before they start. No long lecture, however! Ideally, the lecturer or recitation instructor should have folded the lab into the course sequence. *Somebody* should raise the question, "Why do this lab?" If the lecturer hasn't (for shame), then the recitation instructor should. If not, then the TA should. If all else fails, the student should. (Labs are expensive; the student is paying.) Just in case all else has failed, the TAs should think over this question very seriously in advance, and have some good answers. If you don't know the purpose, or don't know the background of the rest of the course, find out. Talk it over with other TAs. Ask the faculty. Politely, of course.

There's one ingredient about teaching, particularly labs, that I haven't mentioned yet. Enthusiasm. Now it's hard to get enthusiastic about vector addition on a force table. It's like getting a kick out of five-finger exercises for a piano-player. So, hokey it up a bit. Do a preliminary version with three students and three spring scales. (The classic case consists of two big students pulling opposite each other, while a small student pulls at a right angle to the middle.) Besides, as you wander around the room looking at individual calculations from the force table, you'll learn surprising things about your students. Fortunately, it's easier to wax enthusiastic about most introductory physics experiments. They usually involve clever apparatus or surprising phenomena. When the students walk into lab, they should meet an instructor who has the attitude, "Hey, have I got a lab for you!"

It can't always be like that, of course. Sometimes there will be staff problems and sometimes nothing you can do will stir up your students. Even then, and at the very least, this is a golden opportunity for you to learn the freshman physics you haven't thought about since you had to do those labs four years ago. This time, you may understand it.

Any questions before the quiz?

September 1996

Sawing the Lady in Half

As the years go on, we're building up quite an impressive array of physics demonstration apparatus here at Stony Brook. Successive lecturers see something new in the catalogs or invent something to be built in our shops, and the stuff accumulates. We also have state-of-the-art video and computer projection apparatus. When it comes to show and tell, we're big league.

Not me. I still keep a bunch of balls, balloons, string, and Tinkertoys in one corner of my office. Before each class I put a few items in a brown paper shopping bag, to be fished out ostentatiously at the appropriate time. It's a running gag, but with a moral. Physics describes the everyday world as well as the more exotic domains. My message to the students is that with only a few simple objects they too can demonstrate physics, illuminating and impressing roommates, parents, spouses, and other loved ones.

Don't get me wrong. I'm a great fan of magic shows and love to see the lady (always a lady for some reason!) get sawed in half and then reassembled. But in a physics class I want the optics explained. Show and tell should not become entertain and baffle. Furthermore, I love to play with demonstration apparatus, varying conditions and making simple measurements. I like to do it so much that I figure my students will too. Why ruin their enjoyment and experience by handling the apparatus for them? It would be like buying a computer where the salesman tells you, "You can only appreciate this by doing it yourself, but first let me show you a few more things it can do." So I prefer to leave the complicated demonstrations in a corner of the lab room with specific instructions to the students on how and why to play with them.

There are still lots of things left to demonstrate in lecture. At some early time we discuss units. The unit length is the meter. How does that compare with the height of a unit student? Measure one then and there. How big is a cubic meter? That can be demonstrated with a single meter stick and the pantomime of constructing a cube at the front of the room, one meter on a side. What if you filled that volume with water? What would be the mass, and hence what is the density of water in SI units? One? One what? The plausible answer is virtually there in front of you.

Wouldn't a one-cubic-meter tank filled with water be more memorable? Or how about a video showing such a tank being filled with 10-kg buckets of water? Maybe so, but I don't think it's worth the extra effort.

In kinematics, we use algebraic equations and two-dimensional graphs on the blackboard to model some hypothetical reality. Introduce another model—a ball or a toy car. Model the hypothetical motion with algebra, graphs, *and* the solid, three-dimensional object.

Instead of just drawing a diagram of the forces on a book that is lying on a table, place a heavy book on a table consisting of a thin board supported at each end. You can see the distortion of the table and can use a force measurer to calibrate its response. Or make a table with a board and four legs made of springs.

Air tracks and photogates are great for labs, but ball-bearing carts colliding on the lecture table get the point across well enough for lectures. Students with stopwatches can substitute for photogates.

Think what you can do with soap bubbles—pressure, surface tension, light interference. Or Tinkertoys—coordinate systems, statics, moment of inertia, angular momentum. Or a line filament lamp—cylindrical reflection and focusing, slit diffraction and interference. Or rubber bands and ropes and a few force meters—calibrated rubberband force meters, limits of Hooke's law, nature of tension in a rope. How about a hot water bottle, some thermometers, ice cubes, and a heating device for thermo? The students can rub their own hands briskly to convert W into ΔU. When the study turns to electrostatics, an ordinary rubber balloon can pick up a tenth microcoulomb when rubbed on your hair. Two of them hanging from threads fastened to a common point will repel each other. It's a striking sight. The separation angle and distance, and the balloon masses can all be estimated closely, yielding an approximate magnitude for the charge.

Of course there are limits to what you can demonstrate or simulate with simple apparatus. Usually these are the same limits that apply to complicated apparatus, unless that apparatus becomes a project for student handling. Just because the apparatus is simple doesn't mean the demonstration can't be quantitative. Physics should *always* be quantitative. The simpler the apparatus, the more easily you can model the action with numbers measured to appropriate precision and compared with familiar numbers.

Don't confuse the innocent student with fancy technology. Don't think that concepts can be taught without a quantitative structure. In demonstrating physics, *Simplify, Quantify*. If the students can't calculate the pressure exerted by the bed of nails, they've missed the point.

April 1994

The Facts of Life

What a mischievous trick nature and society play on you when you're young. During the disturbing and wonderful days of adolescence you are required to make decisions and take actions that shape all the rest of your lives. It should be a time for worried and secret and passionate exploration of yourself and of your place among family and friends. It should be a time for dreaming and contradictions and defiance and love. It must also be a time for getting good marks in school, on SAT exams, and in the right courses.

There is almost no second chance. Either you study and do well in math and science courses in high school or you have closed the doors leading to a wide range of professions. Without algebra, trig, algebra, geometry, and yet more algebra, without physics and chemistry and biology—but especially physics—you stand a poor chance of doing well in college physics. Without that introductory course in college physics, best taken in the freshman year, you are blocked from advanced courses in physics, chemistry, engineering, astronomy, earth sciences, and life sciences. Physics is required by most programs in any of the quantitative disciplines.

Many colleges still hesitate to recommend specific high-school courses as prerequisites for admission. They prefer to admit students to the college in general, and not to any specific major. Some even refuse to let you choose a major until the sophomore or junior year. For the fields where physics is required, the colleges are kidding you. Without physics in the first year, or the second at the latest, and without math continuously, the required sequences for scientific and engineering professions become impossible. Without good high school preparation, your success in the college courses is seriously jeopardized.

There you are, sweet sixteen, and the doors are already slamming shut. They don't have to. Not if you, and parents, and counselors, and teachers are aware of the consequences. The doors should remain open for all students—males, females, black, white, rich, poor. With math and physics in high school and your freshman year in college you can go on to major in physics, chemistry, geology and all the other physical sciences. You can become an engineer or study computers. You can become a teacher of physical sciences. You can go on to pre-med training or any of the other life sciences. If you change your mind, nothing has been lost. You will still have room for all the required sequences and electives in any other career, from home economics to comparative theology. But if you postpone the study of physics, if you don't make adult choices until your third or fourth year of college, it's too late. The required sequences for the technical fields take several years, and they start with physics.

Of course, physics can be fun, and physics can be cultural, and people ought to take it because it's good for them. But physics also opens doors to a whole range of professions and future employment. Keep the doors open.

December 1982

The Text For Today

And further, by these, my son, be admonished: of making many books there is no end.　　　　　　　　　　　　　—Ecclesiastes 12:12

What a wealth of resources we physics teachers now have available! In this issue we celebrate high-school physics texts. There are many to choose from and the quality is high. Compared with the pre-Physical Science Study Committee (PSSC) days our current texts are mostly free from factual mistakes and are generally written by people with a good understanding of modern physics. Thirty years ago this was not the case. I remember displaying the most popular physics text at that time as a horrible example of what had gone awry in science education. The most advertised feature of the book was a four-color transparency overlay showing a steam shovel in a pit. Take off the top transparency and you could look in the cab. Take off the next and you could see the gears. So much for modern physics! Years later when I chided the editor responsible for the book, he chuckled and said, "Wasn't it frightful? But it swept the nation."

Thirty years ago, following the lead of the PSSC, all publishers transformed their texts. It was one of the major by-products and accomplishments of the PSSC project. So now we have reasonably accurate texts to which authors and reviewers have devoted years of writing and checking, on which publishers have gambled huge sums for production, and for which school boards and taxpayers have sacrificed their money. So who reads textbooks? Not students, apparently. I seldom find an entering college student who knows the name or author of his high-school physics text. In fact, the most common claim is that they didn't use a text in high school, except maybe for homework problems. In one school I'm familiar with, the teachers themselves boast that their students don't need the texts on their shelves. They have their own program. In another school the teacher says that most of today's texts are too difficult for today's students.

Now it is a truism that no one is satisfied with any particular text, even the author. What is more natural, then, than to supplement a text with your own material? In fact, if you accumulate enough of these sections, you have a start on another text and should be sounding out publishers. Usually, however, these home-produced texts have serious drawbacks. I have seen many, but have seldom seen a good one. The writing of most of us needs reviewing and editing. Amateur composition and reproduction is usually hard to read. But even if your writing is lucid, your physics sound, and your typography professional, what makes you think the students will read your opus? They appear not to read anything. At every level we hear the complaint that the text is either too difficult or is not needed. At most, the student thumbs through the chapter to find the right formula in order to do the homework problem.

In a junior-senior level college physics course we are using one of the classic texts. My students say that it is too terse and sparse for them to read. After exploring how they use physics texts I ran a remedial reading session for the whole class. First we scanned the chapter, and then went back to the first section, questions and pencils in hand. Of course you must have paper and pencil while reading a physics text and of course you must have questions. At every step you must fight back, checking up on the author, following through on every statement that "it can be shown" or that "as we saw on page 132." With publicly owned texts, the students should not write in the margins, but they can and should produce their own text notebooks of criticism and commentary.

Years ago the first edition of a popular text in introductory quantum mechanics had a number of typos and physics mistakes. According to legend, the author responded to people who wrote him about these by saying "Thank you, and congratulations. You're right and I was wrong. I'll be sure to keep that mistake in the next printing so that other readers can be similarly tested." What a great idea! What we need are texts with a carefully planted mistake on every page, preferably in some formula needed for homework. If the kids don't find the mistake they get their homework wrong. The book might be called *Phallacious Physics.* That would really be educational. Everyone should grow up with a deep love for the printed word and a deeply ingrained skepticism about its truth.

As for students not needing to read texts because we teach so well, maybe we're doing them a disservice. Maybe it's our obligation to teach our students how to read physics, which is a different skill from reading the *Hardy Boys.* Then it might not be necessary for college students to buy study guides on *How to Read the Text,* or the cartoon sequels on *How to Read the Study Guide.*

In the meantime, authors are still confident and writing, publishers are still hopeful and printing, schools are still trusting and buying, and of the making of many books there will surely be no end.

November 1982

Y ou all know about the child who came home from Sunday School and told her parents that they had learned a song about Gladly, the cross-eyed bear. For years the *Reader's Digest* has been telling us about the cute misinterpretations of children. Also, for years, John Renner, Anton Lawson, and others have been telling us that many of our high-school physics students really can't do proportional reasoning, among other things. Neither can a lot of our college physics students.

Twice in recent months I had my nose rubbed in these disagreeable realities. In late spring I wrote and administered a questionnaire to children in a local Sunday School. We wanted to find out what they knew about God, Jesus, the Bible, and their church. Don't be concerned about the religious aspects of this story; we were asking where, when, what, who questions. Younger children were interviewed, older children wrote their comments and answers. Of course the answers revealed a progression of ability and understanding from first grade to high school. What was surprising was that almost all the teachers were surprised—if not dismayed. It appeared that the children had absorbed very little and it also seemed obvious that the texts were wildly mismatched in both reading and conceptual level.

During the summer here at Stony Brook we give a short multiple-choice quiz to incoming freshmen during orientation sessions. The papers are machine graded and the results become immediately available. Several of our experienced faculty had agreed that unless a student got at least five out of ten correct on the quiz used this summer, the student should sign up for an interview. After checking SAT and high-school grades, the counselor then might bully the kid into taking our one semester Prelude to Physics course. (It's a holding pattern that teaches them, among other things, how to hold a meter stick and how to draw graphs. It also gives them a chance to grow up.) Unfortunately for the counselor, only about 2% of the incoming freshmen got five or more correct.

The quiz was designed to test several different capabilities. We should not have been surprised, I suppose, that practically no student could figure out the following problem: If you want chopped apple for a salad, about how much more will you get from an apple with four-inch diameter than from an apple with two-inch diameter? a) twice, b) four times, c) six times, d) eight times, e) sixteen times. I was personally surprised, however, that we also got poor scores on two straight physics problems, with very simple arithmetic. One was to find a, given F and m, and the other was a simple series Ohm's law problem. Most of the students taking this exam had just finished high school physics.

But we all know these things and really shouldn't be surprised. It's common folk wisdom in our profession that incoming students give no indication that they have ever studied science before, and, of course, it appears that they have had no math at all. Nevertheless, we and the textbooks go hurtling off, assuming that all our students not only like vectors but can easily do the algebraic manipulation of the trigonometry. With these assumptions, we here at Stony Brook, like many other large universities, manage to shake off about 40% of our students by the end of the year. A large fraction of the survivors, as we know very well, do not quite fail because they learn rote procedures and answer patterns in lieu of the physics.

And do you know your new students? Just for the masochistic fun of it, this fall why not have them write answers to questions such as these:

What's the center-to-center distance between your eyes? a) 1 cm, b) 65 mm, c) 140 mm, d) 2.5 cm, e) 1.5×10^{-2} m.

A friend is riding past you on a bike. A book falls off the pack rack. Draw a diagram and describe in words the path of the book as it falls. What forces are acting on the book as it falls?

What's the phase of the moon during a total eclipse of the sun? Illustrate with a diagram.

Sketch a graph of the sine of an angle versus the angle as θ goes from $0°$ to $360°$.

These questions may be too simple for your students (though not for mine). In making your own diagnostic quiz, go over the first few weeks of your course and ask what you will assume that the students already know. At least some of the questions should require short essay answers or explanations. Who knows, you may discover that all your students are in Piaget's formal operational stage, and that they remember everything taught them in algebra, trig, earth science, and chemistry. Then you'll be all set to brief them on vector operations and hurry on to the good stuff in dynamics. Or the results on the quiz may lead you to a slightly different conclusion. You may want to temper the wind to the shorn lamb. In either case, it's a wise teacher who knows his own students.

September 1984

How to Win the Science Project Contest

When I was in seventh grade I bought two 6-in glass disks, some abrasive powders, and a book on amateur telescope making. After I had the mirrors ground and pretty well polished, my father took me to see a local machinist who had made his own telescope. His mounting system was handsome and far beyond the cardboard and orange-crate arrangement that I had in mind. This man gave me some good advice. He said that I ought to work very hard, because soon I would discover girls and then I'd give up. Alas, the advice came too late. I never did finish the telescope.

Here at Stony Brook we get lots of queries from Long Island students about science projects. Everyone wants to win the Westinghouse, or at least get first prize at the school science fair. You must get such questions, too, and no doubt have your own suggestions for students. Of course, the best advice any of us can give is for the student to select parents who work in university or national laboratories. For some reason, major prizes frequently go to such students. Just natural talent, I suppose. Short of that, here are some other things we suggest.

Start your project with a feasibility study. Real scientists do. Can you imagine walking into Brookhaven National Laboratory and saying, ''Today I'd like to use your Alternating Gradient Synchrotron to bombard either lobsters or silicon diodes to see what happens to them.'' Say what? What's the question? What prior research have you or others done? What skills do you have in this field? How much beam time do you need? What energy? How much time can you spend on this? How much help do you need and is it available? What apparatus do you need and is it available? How much will it cost? Feasibility studies are not unique to science projects. We all have to do feasibility studies in everyday life. You can't go grocery shopping without a list of needs and a rough idea of the amount of money in your pocket.

The feasibility study should be about one page, typed, double-spaced, probably with diagrams. Use English sentences. Make sure you spell out your estimate of the required time and money, and your ability and commitment to meet both. How many evenings? How many weekends?

Research results must be quantitative. This doctrine is one of the hardest to follow, but without quantitative analysis the project is not all that, you know, like, great. For instance, last year at a science fair I looked over four solar-energy projects. Each student had designed apparatus to capture or concentrate sunlight for heating or photoelectric use. Not one knew the average solar energy per square meter per second at our location.

Science involves numbers and magnitudes and quantitative relationships. Be sure you have error bars on your graphs. (Don't know about error bars? What a great science project it would be to learn and illustrate their use and importance.)

When students ask for suggestions about possible projects, we should have two lists. First, there should be a list of projects not to tackle. Don't try to do subatomic physics, either experimentally or theoretically. Even if your mother has a collection of particle track photos from her earlier research, you can't really find anything new during your junior year in high school. Leave string theory to people who have nothing better to do. Don't plan to put living things (even spores or invertebrates) in zero g to see what happens to them. You can't produce zero g for more than a fraction of a second in apparatus easily available here on earth. No way. Don't build scale models of anything and expect to investigate the dynamics of full-size devices. No wind tunnels. At least, don't do this unless you understand scaling principles. On the other hand, the investigation of scaling laws might make a very good project.

What can a high-school student reasonably do for a real science investigation? Don't try to make a giant leap. Instead, attempt something just one step beyond your school work. For instance, why not produce rainbows or similar natural optical phenomena? One of the first experimental problems is to keep water from running all over the floor. Beyond that, the physics can be approached, understood, measured, and photographed in several stages of approximation. There's the simple introductory textbook theory, and then there are more sophisticated effects. Besides, the effects are pretty, and it's very satisfying to produce a rainbow. (After your first rainbow, try for the "glory.") Do you know that very few people know how rolling friction takes place?

Most textbooks ignore the problem or have it almost wrong. Why not rig a bike wheel so that you can take pictures of what happens during braking or acceleration? Then explain how the retarding friction is different from the traction, and why neither has anything to do with the coefficient of static friction. This is a tough one. You'll need lots of advice. Do you know how crystal radios work? Not many people do. What electric-field strength can be detected with such a set? Over what voltage range can a solid-state diode rectify? How do you get that voltage across the diode? How much energy is needed to actuate earphones? How can you extract that much energy out of space? Do you know how intense the 60-Hz fields are around your home appliances? A lot of people are worried about that problem these days. Why not make an instrument that can get quantitative readings? Do you know how intense various sounds are in your home or school? Why not measure those intensities, and perhaps find reverberation times for classrooms and auditoriums?

The lists of possible and impossible projects will vary from one locality to another. If we prepare such lists, we should make sure that we have a rough idea of why the projects are or are not feasible. There are just two other points about scientific research that we can tell our students, but they can really find out only by themselves. Most of research is drudgery and hard work. Pipes leak. Circuits burn out. It's hard to keep notes and write reports. On the other hand, sometimes things work. Occasionally you find something that no one knew before. But that doesn't happen often in high school. More likely you may suddenly realize that you understand something you hadn't known before. Then you know the thrill of research. It's a high unlike anything else. No matter how the judging turns out, you've already won the science project contest.

January 1990

Judgment Day

I n spite of the injunction in Matthew 7:1, you and I are paid to judge. It is generally thought that our main job is to instruct. To the extent that we have received any training in our profession, besides the subject matter, the training has mostly been concerned with methods of making the subject matter understandable. Most of the pages of this journal are devoted to that end. Nevertheless, society demands more of us. In that awful day of judgment, we must grade our students.

Since final exam time will soon be upon us, it seems to be an appropriate season to ponder exam techniques. Actually, we should have started considering the subject last summer. The time to make up the final exam is before classes begin. *Then throughout the year you can teach to the exam.* Is that a scandalous suggestion? Let's be practical. Of course we think it necessary and appropriate to have goals for the year's work. These are spelled out in terms of a syllabus, either homemade or state-mandated. The goals are matched to the text, or vice versa. Out of the text we select certain homework problems and laboratory exercises. Thus our goals are manifold and manifest. Or so we think.

What about the student goals? Whether in high school or college, students are very concerned, if not primarily concerned, about their final grade. They want to study and memorize whatever is going to be on the exam. One of the great selling points of any college fraternity is its library of past exams. In New York State, you can buy review books of previous high-school Regents exams. These become the supplementary texts in high-school classes at least during the last month of school. No matter what the state syllabus says in its carefully worded generalities, this year's final exam is next year's syllabus. Any good student knows that.

Well then, let's make up our exams before we start the course, and let's make sure that the exams test whatever goals we are attempting. If we want our students to memorize the arcane vocabulary of our profession, including the Greek symbols, then the testing is easy. Multiple-choice tests are ideal for the purpose. If you're going to work in physics, you have to know the language. No doubt that ability should be tested early and late. The same goes for other parts of the language that we use: *Work has the same units as (a) force, (b) momentum, (c) energy, (d) moment of inertia.* Does it sound dull? French vocabulary drill is also dull. It has to be tested, but if that's the main part of the exam and thus the main part of the course, the students will hate French and never get to use it.

If we want our students to be able to use physics, then the exam must test that ability. That's very hard to do with multiple-choice questions. Recently a number of teachers got together with me to critique an exam in physics. This particular exam required us to read and analyze some paragraphs or diagrams before answering the multiple-choice questions. None of us could do well in the time allotted. But quite a few students had done very well. They knew how to take multiple-choice exams. Most questions can be answered by inspection, or at least the choices can be narrowed. Then, if necessary, go to whatever reading or diagram is provided. That's the way most students do their homework. Instead of reading and analyzing the chapter, they turn immediately to the problems, thumbing back through the text for equations that have the required variables.

It may seem that it is impractical to test large numbers of students with anything but a machine-gradable exam. There are very real problems, of course, but consider these two exam questions. The first is from the New York State physics Regents exam of 1917: *Describe with the aid of a labeled diagram the essential parts of an ammeter. Draw a labeled diagram showing how the instruments should be connected in a circuit to find the resistance of a given wire by the voltmeter and ammeter method.* In the 1988 Regents there is a diagram showing a 110-V source in a series with three resistors. Voltmeters are shown across each resistor. One reads 20 V, another reads 20 V, and the problem is: What does the third one read? The choices are (a) 20 V, (b) 70 V, (c) 90 V, (d) 110 V. Those two questions, 71 years apart, are testing for very different abilities and surely reflect very different courses and teaching.

In most cases, the testing instrument doesn't make too much difference for grading purposes. For various reasons we are required to rank our students. Good students, as determined by our informal observations, usually do well on written exams of any type. Furthermore, most exams are reliable, because the good students do uniformly well and the poor students do uniformly poorly. But not always! Once I had a class where the star was very skilled at exam taking, but I was sure that he had less creativity and weaker insight than another student who rarely did well on the formal exams. In the years since, the student who got all the scholarships has never accomplished much, while the slow but creative student has had a successful research career. The uniform, machine-gradable exam is particularly unreliable in judging the unusual cases.

Besides reliability in exams, we require, or assume, validity. Validity is not just the quality of being technically correct. A valid exam should test the skills and knowledge and ability that we have been trying to instill in our students. That's why we should make up our exams in advance. A sample of everything we teach should be tested in some form. If we expect our students to be able to wire a circuit, or to set up a lens system, then we should test for that skill. If we ourselves are fascinated by the accomplishments of our science and are still thrilled by the mysteries remaining, and if we have tried to communicate that to our students, surely at the end we should find out whether they got the message.

Tests can be a plan for ourselves as to what we teach. They can be a prod and a guide to our students about what they must learn. They can be instruments of judgment for ranking our students and predicting their future performance. Tests and their results can also be revealing to others about how we teach and what we teach. Thus we should bear in mind the flip side of Matthew 7:1. Inevitably, as we judge, which we are paid to do, we also are judged.

May 1990

Judicious Hoopla

I t is well known in carnival circles that you can't sell saltwater taffy unless you first get the customers into the tent. That's what barkers are for. Similarly, you can't teach Newton's laws to students who aren't in the physics class. Somehow, we have to make the subject attractive and vital to our prospective audience. We also have to hold their attention once they're in. One way to accomplish both missions is to use lots of demonstrations.

It's always fun to see good demonstrations, particularly when they work. The physics lecturer in my freshman year was demonstration happy. He had an assistant to set things up in advance, so that all the professor had to do was press buttons or pull triggers. One day the professor showed us that all-time favorite, The Monkey and the Hunter. At the start of class he pointed out the can fastened to an electromagnet on the ceiling and the spring gun on the table aimed directly at the can. He fired. He missed. He got the bamboo pole, repositioned the can, aimed carefully, and missed again. Half an hour and many tries later the pellet hit the can in midair. The jeers turned to thunderous applause and we were dismissed without further explanation. I resolved then and there that in all my future career I would never try to figure out the algebra and geometry of that situation. (Now that I know about reference frames, there's no need to write down the equations. In the reference frame of the falling monkey, the result is terrifyingly obvious, and the solution is trivial.)

Ever since that freshman class I've been a little jaundiced about physics demonstrations. Nevertheless, I use them all the time. As Wordsworth once said about something else, "To the solid ground of Nature trusts the mind which builds for aye" (pronounced ā). That's a great slogan for physics teachers. We have to show our students that the symbols we use actually describe a solid world. However, we must make sure that the solid world is not more mysterious than the symbols. We didn't have the magic of television in my freshman year, but we did have magic shows. A friend of mine could pull coins out of people's ears and I had seen ladies sawn in half. So what's so wonderful about having a ball hit a can in midair? Why shouldn't the ball hit the can? Before we explain the wonder we must make sure that the students do indeed wonder. The demonstrations themselves usually need a lot of explanation, followed by calculations and physics principles. Without these explanations, you might just as well saw a lady in half.

There are other ways to get people into our tent. Good science museums try to lure the spectator into becoming a participant. Following the lead of the Exploratorium, modern science museums have hands-on exhibits. The visitor can control some variables and even make measurements. Once again, however, without preparation and follow-through, most of the visits are just entertainment. It is particularly discouraging to see most school classes at such a museum. Usually the kids just wander around pressing buttons and waiting for the explosions. The main benefit seems to be that they get a day off from school. The visit can be educational only if the teacher knows in advance what will be on display and assigns specific measurements to be made and reported.

Other ways to make physics fun are amusement park visits or a physics olympics. These are particularly good for press coverage, with lots of photo opportunities at the roller coaster or the egg toss. Do these projects increase enrollment? The evidence is poor. Perhaps they're good for class morale. Do they increase understanding of physics? Only if each activity becomes a subject of instruction or student research, preferably involving measurements and calculations.

Seeing is not necessarily believing, and seeing, by itself, seldom leads to understanding. Good demonstrations can be spectacular, but their operation must be simple and transparent. The performance should be preceded with an explanation of the mechanism by the instructor and speculation by the students about the outcome. There probably isn't time for more than one demonstration per class. It takes a while to raise the question, explain the apparatus, explore predictions about the outcome, make the measurements, and then compare the quantitative results with theory or expectations. Really good demonstrations will be quantitative, not with fussy measurements but with appropriate approximations. How many newtons were exerted on how many kilograms? Was the observed travel time in seconds approximately what you would expect? Without numbers our science is weak and leaves our students powerless.

I love theater and I'd hate to teach a physics class without a demonstration. I also like to visit science museums and amusement parks and I have a good time at the annual physics olympics. It's fair game to use all these activities to get students into our tent. If it helps enrollment, we can even hand out balloons and free prizes and hints of forbidden pleasures. We know that inside the tent the prizes and pleasures are real but require work. Let's not diminish the real performance by substituting the hoopla we use at the entrance.

September 1990

To the Solid Ground of Nature—

For over one hundred years, the cover of each issue of the English journal *Nature* carried the admonition of William Wordsworth, "To the solid ground of nature, trusts the Mind that builds for aye." Before taking these words to heart, you must realize that in this context *aye* is pronounced *āye,* an archaic term meaning *ever* or *always,* as opposed to *eye* meaning *yes.* Since all right-thinking physics teachers will agree with these words for aye, let us now consider their application to our instructional labs.

It is a truism that physics is an experimental science, and it would seem to follow that in order to learn physics you must study laboratory technique. Actually, we know it doesn't work that way. Many famous theoretical physicists spent precious little time in the lab, either during school days or afterwards. Wolfgang Pauli was notorious for the baleful effects that he had on any lab equipment in his vicinity. On the other hand, there are many good examples of how the great innovators thoroughly understood the experimental work that others were doing and had a perceptive sense of its relative importance and validity. For instance, Yang and Lee based their conjecture about parity violation on their detailed analysis of experimental reports. If good physics can be done, or learned, without going near a lab, why attach lab sessions to the introductory course?

We raised this issue long ago in the January 1979 issue of *The Physics Teacher* (*TPT*). In the editorial I listed, tongue in cheek, all the reasons for not having the introductory physics labs. For instance, students frequently get incorrect results, and so learn to lie, or round off. Also, abandoning a double lab period would immediately reduce teacher work load by 16.667%, thus allowing the teacher to do six sections instead of five, or be given cafeteria assignment.

Fearing that some dean or vice principal might get hold of my list, I ran a corrective editorial in the March 1979 issue. Labs are good. Cross my heart! You know it and I know it, and it's just a shame that we can't prove it. As far as I know, there is no valid evidence that the standard, traditional introductory physics labs have any effect on the results of standard final exams. Even exam questions on experimental results or the details of apparatus can be best answered by reviewing books or other exams. The actual performance of the experiment doesn't seem to be important.

It's a complicated problem, with a long history, but the standard lab with its directions that no student reads in advance, its time-limited series of steps, its apparatus that frequently breaks, its requirement of a formal, written report—it just doesn't work well. Students ought to enjoy lab, in the same way they enjoy band rehearsal or drawing class. They should be out of their seats experiencing the actions and surprises of nature in ways that aren't possible with lectures or reading. We have some evidence from researchers such as Richard Hake (*TPT* 30, 546 [December 1992]) that some nontraditional techniques of lab instruction can make a difference in student performance on exams. Maybe it's time to break the mold.

Why not separate the two main functions of introductory physics labs? The first is the provision of apparatus and occasion for "handling the phenomena," of experiencing the solid world of nature. The second is the training in skills of measurement, procedures, and reporting, and the experience of tackling a research project. The type of skills depends on the nature and level of the course, but you surely can't teach the scientific method in a series of one- or two-hour periods. (Remember Percy Bridgman's definition: "The scientific method consists of doing your damnedest to understand nature, no holds barred.")

The weekly lab should be a pleasant adventure, with plenty of advance hype by the instructor and anticipation by the student. There should be several phenomena to be handled and manipulated, with order-of-magnitude quantitative analysis. In style the lab might be a small-scale equivalent of going to an amusement park, with some simple measurements to be made and a minimum amount of reporting to be done. But the lessons before the lab should instill curiosity and interest about what may happen, and the lesson after the lab should follow up with discussion and analysis to make the experience relevant.

What about the second role of laboratory instruction? Let the students do one or two real experiments each semester. Propose a list of possible projects, appropriate to the level and type of class. Encourage students to make up their own project if they wish, subject to your approval. Choose problems that can be done mostly at home or in the dormitories. The first requirement should be a one-page feasibility study, to be approved before they proceed. What do they propose to find out, and to what precision (order-of-magnitude? factor of 2? 10%?)? What equipment will they use and where will they get it? How long will it take? Most students have never made out a shopping list before, to say nothing of a feasibility study. Writing a proposal is the very essence of the scientific method. The project itself should take some weeks and should be described in a fairly formal report.

In choosing phenomena to be handled, whether in the weekly lab or in the semester projects, I would suggest some guiding aphorisms:

1. Never simulate, whether by film or tape or computer, any phenomenon that you can produce in the real world.
2. Demonstrations are good; student handling of phenomena is better.
3. Concepts are best understood by using your muscles; the larger the muscle, the better the understanding.
4. Finally—to the solid ground of nature, trusts the Mind that builds for aye.

October 1993

Physics Teaching and the Doctrine of Invincible Ignorance

O ne of the ancient problems of Christian theology concerns the fate of those who, for one reason or another, have not received the Good Word of salvation. It's clear that hellfire awaits those who have heard, understood, and deliberately ignored the message. Presumably the patriarchs and other worthy ancients got the Word and were saved during the three days that Jesus spent in hell. There is also partial salvation for those whom the missionaries never reach. You wouldn't want to damn these pagans outright, but on the other hand you wouldn't want them mixing with the heavenly throng. They are assigned to purgatory where at least they will not be uncomfortable. What do you do, however, with the unwashed heathen who do get to hear the lectures, but fail to believe through sheer stupidity? To save these souls from damnation, the Church allowed them into purgatory through the Doctrine of Invincible Ignorance.

Every physics teacher recognizes this situation. We have all known students who try hard but just can't understand physics. In recent years that common observation has been raised to the level of a doctrine through the work of the Swiss psychologist, Jean Piaget.

Piaget has been studying young children since the 1920s, and has developed a sophisticated theory of learning. The part of his observations and theory that has recently been popularized concerns developmental levels of mental growth. We have carried articles on this subject by Renner.[1] A child advances from the stage of sensory-motor to preoperational to concrete operational to formal operational. Since this final stage involves the ability to deal with abstract operations such as algebra, its attainment would seem to be a prime requisite for the study of science, particularly physics. In the usual age classification of the Piagetian levels, youngsters enter the formal operational stage at puberty. Presumably that circumstance matches well with various traditional rites celebrating arrival at the age of reason—church confirmation, bar mitzvah, junior high school, etc. Certainly it would seem that

1. J. W. Renner and A. E. Lawson, *The Physics Teacher*, 11, 165 (1973) and 11, 273 (1973).

by the age of 16 or 17, every youngster should be well into the formal operational stage and thus be capable of learning physics. Failure of such a student to understand physics must then be ascribed to laziness or other forms of wickedness.

We all know, however, that a lot of students never seem to make it into the formal operational stage. Renner quantified this common knowledge and at an annual meeting of the American Association of Physics Teachers (AAPT) in Anaheim there was a workshop on the subject. Under the chairmanship of Bob Karplus, a committee consisting of Bob Fuller, Frank Collea, John Renner and Les Paldy produced an introduction to Piaget as applied to physics teaching and provided actual examples of the problems.

It turns out that lots of students, even those doing well in other subjects, do not really comprehend the concepts of functional dependence, or proportionality, or scale, or graphical representation. Without these concepts, can anything useful about science really be taught? Lots of schools and colleges apparently believe so because they give physical science courses that require no formal operations. We are reliably informed by several publishers that science texts using algebra will not sell for courses for future elementary school teachers. Our own prejudice is that such a course is worthwhile only if it is taught at the concrete operational stage with actual manipulation of phenomena. It is harmful, however, if it becomes one more lecture course in rote memorization of science words and classification schemes.

How do you know whether or not a student is in the formal operational stage? The tests and research are primitive and human growth is complex. Perhaps a student can't do algebra with symbols but can think logically about sequential events and dependencies. Perhaps a student is on the borderline of abstract understanding and only needs experience.

For that matter, are we sure that there really are conceptual levels, determined primarily by chronological growth? Granted that many or most eighteen-year-olds cannot use simple algebra, can they not now be taught the principles and so be admitted to the greater glories of our scientific age? Maybe not. Maybe most people can never enter the formal operational stage. No one knows enough about human growth and learning to answer this question. We would not attempt to teach algebra to a three-year-old, and we accept the fact that some people just can't learn to carry a tune.

At the very least, we in the education business should become more knowledgeable about the subject. We should already know enough to avoid trying to accelerate schooling by introducing abstract concepts before students have experienced concrete phenomena. Meanwhile, if we give up the attempt to teach physics to everyone in this scientific age, we acquiesce in a form of secular damnation. On the other hand, maybe both students and physics teachers would be happier if we accepted the conclusion that for many students certain types of ignorance are invincible.

March 1975

On Teaching That We Do Not Know

F all came early this year. Somehow I wasn't expecting it for several more months. When it became clear that summer was over and that school would really start, I began to prepare my annual outline of topics, facts, and skills that I would teach during the year. The list is not quite the same as last year's because I want to try a different emphasis in mechanics, and there are some interesting developments in optics. Needless to say, the list is a summary of things I already knew or have just learned. You can't teach something you don't know!

The trouble is, there's a lot about physics that I don't know, and I discover new things all the time. Great vistas of ignorance keep opening up. In fact, I now know so many things that I don't know that it's not clear that there's been any overall gain. Let's face it: in the great competency-based exams of life, I have not achieved mastery.

This realization of my limited knowledge would make me feel worse if I didn't live in a large and powerful university physics department. At least once a week a journal or a manuscript arrives with a fact that I hadn't heard of before or an argument that doesn't seem quite right. So I'm stupid, but at least can't I get expert advice just down the hall? How frustrating, but how satisfying, it is to discover that in many cases my colleagues are also baffled. Ignorance, it turns out, is endemic among physicists. It's also a lot of fun. The blackboards around here are covered with partial calculations and tentative sketches left over from bull sessions about things that we aren't sure of. If we already knew everything about the subject, we'd be bored stiff.

But should we let the students in on this? I was in a high-school classroom some years ago, and heard a student ask the teacher a very perceptive question. A look of puzzlement and then dismay swept over the teacher's face. How embarrassing! "I don't know," he said. "I've never thought of that question before. How will we find out the answer?" It takes a master teacher to do that. Maybe we should all learn how to admit our ignorance. In fact, maybe the most important thing that we can teach about physics is that there's a lot about it we don't know.

Let's list the categories of our ignorance. First of all, science is surrounded by questions that have been bypassed. Instead of asking why an object should keep moving, Galileo asked why it should stop. Instead of worrying whether light is a particle or a wave, we now require only that our equations predict successfully the results of experiments with light. Of course young students keep asking the old dead-end questions. Somehow we must let them discover the power of asking the right question, without squelching their annoying tendency to ask questions in a form for which we have no prepared answers. They must learn that physics doesn't tell why; it tells how. And then they ought to mistrust that philosophy enough so that they occasionally keep on asking "why?"

The most embarrassing category of ignorance consists of those facts or skills that are really part of our teaching repertoire but which we have forgotten or never learned. I just found out last month how to determine the radius of percussion of a baseball bat. No excuse. Somehow I had missed that in school and in all these years had never wondered about it before. I suspect that it might be reassuring to a student to find that his teacher doesn't know all the answers right offhand, even to simple things. Besides being reassuring, it would also be instructive if the student could then work with the teacher and watch his technique as he tries to find the answer.

There's another realm of lack of knowledge for most of us. If we specialize in subatomic particles, we probably aren't up-to-date in solid state physics. If we are skilled with the equipment and techniques of experimental physics, we may have trouble with the manipulation of certain mathematical forms. If you think a professor of theoretical physics has no blind spots, invite him to teach anything to a high school class for a few weeks. Students ought to know that everyone has blind spots. If one appears in you make sure that the class knows its depths and boundaries. That doesn't mean that we should be proud of our ignorance, like the English teacher who says, "Oh, I just can't understand science." On the contrary, let the students know that we can be fascinated by any topic that is of interest to some other human. We may not have studied that field so far; we may not have time now; but if the student is interested, how can he find out? The skill of the physics teacher should be not so much in knowing things, as in knowing how to find out.

Finally, there's one great truth that we must make sure our students understand. Besides the facts that we don't personally know because we are forgetful, or stupid, or limited, we are surrounded by mysteries to which no one knows the answers. That's what physics is all about. What a dull field this would be if it consisted of a closed bunch of facts and techniques. No one knows the complete nature of the subatomic particles. No one knows (in spite of the August discovery) whether or not there is a magnetic monopole. No one knows whether the universe will expand forever. No one even knows what humans are doing in this universe. The science of this century hasn't explained everything; instead, it has raised the curtains on regions that we never knew we didn't know. The physical universe is much more mysterious than it was in 1900.

The only excuse that I have for listing the things that I know, and which I will teach this year, is that it's a short list. I don't even know how long the list would be of the things that I don't know. But that's the list that makes the teaching enterprise forever fresh and fascinating. This year I intend to make sure that my students know the awful truth about how little I know, and the exciting truth about how little any of us knows.

October 1975

What a Young Man's Fancy Does Not Turn to in the Spring

What period of life can compare with that senior year in high school? There are the continual placement exams, the admission forms and interviews, and then the breathtaking wait for acceptances. There are parties, and yearbooks, and senior plays, and senior dances. There is also a physics course.

I remember the friendly advice that I received some years ago, just before starting to teach the new Physical Science Study Committee (PSSC) course at our local high school. "Finish the course before Easter. After that there is only time for the senior trip, the play, and then the preparations for the graduation dance." Naive as I was, the advice seemed all wrong. After all, this class was made up of the best students in the school, and the subject was intrinsically fascinating. By spring I knew that my advisers were right. The kids were under terrific pressure from the whole community to take an active lead in the social and ceremonial affairs of the school. These were bright youngsters and they lived up to their public responsibilities. First things first! By that time I quite agreed with them; compared with these other activities, physics was very unimportant. Besides, these students had already been accepted by good colleges. Many of them would take physics there. It is well known that there is no good evidence that "taking" high-school physics helps in the study of college physics.

In some communities the trilogy of science courses is presented to upper-track students on an accelerated schedule. Biology, chem, physics in ninth, tenth, eleventh. The senior year is then free for fun and games or a college Advanced Placement (AP) course. The AP physics course is, in many schools, a travesty, but who can quarrel with fun and games? The trouble with the accelerated program is not that it leads to twelfth grade foolishness or idleness, but that it accentuates the illogic of teaching bio before chem, and chem before physics. The new biology courses really require some understanding of chemistry and physics, and, of course, CHEM study is mostly what we used to call physical chemistry. Lacking a rational sequence, at least we should present the courses to students who are as mature as possible. Even the bright students in ninth grade are not mature enough to appreciate a course like Biological Sciences Curriculum Study (BSCS).

Some day, now that the twentieth century is here, our profession must really face up to the problem of logical subject development. The problems, of course, are enormous. There are all those textbooks. There are also all those biology teachers—largely because there are all those ninth and tenth graders who have to take one science course. Until the millennium comes, we ought to realize that it is only in junior high school that most students see chemistry and physics. The physics teacher might exploit this situation by becoming personally involved in what goes on in the junior high. There are some good courses available these days—several of which will be described in forthcoming issues of *The Physics Teacher*. There is the Introductory Physical Science (IPS) course for ninth graders, a descendant of PSSC that concentrates on the theme of atomicity. There is Time, Space, and Matter for eighth graders, a delightful mix of physics, chemistry, astronomy, and geology. There is the Intermediate Science Curriculum Study (ISCS), a coordinated seventh, eighth, and ninth program, designed for individualized study, and starting out with a mix of physics and chemistry.

There may come a time when the formal physics survey course for twelfth graders has no customers. A lot of the fundamentals of physics can be learned in the junior high years, especially those concepts that are learned through the hands instead of with symbols. The tenth and eleventh grades might be devoted to sequences in physical and biological sciences, in that order. The senior year could then be reserved for project type courses in any of the sciences. (Reserved only up until Easter, of course. Who would want the spring of the senior year to be devoted to anything other than causes, festivals, parties, and love?)

January 1970

The Hawthorne Effect and Other Mysteries

And then he asked The Question. "How did you prove," he said, "that your new program is better than the traditional?" What a reasonable question to ask of any educational innovation! Why not give a scientific answer? Choose an experimental sample of students, a closely matched control group, and then test them.

That's just what one group of educational researchers did years ago in order to judge Physical Science Study Committee (PSSC).[1] With the full ritual of statistical jargon they normalized their two samples of less than one hundred each to five significant figures. Apparently still serious, they then used as the single criterion of success a test that had been designed for traditional physics. Naturally, the control group taking the traditional course did slightly better. Ergo, traditional physics is better than PSSC physics!

My questioner, it turned out, wanted me to test a grade school science program by seeing whether or not the students knew "their facts." What facts? Why, the facts a student ought to know about science at the end of sixth grade. The names of the planets, perhaps, or the fact that ". . . . the sum of energy and matter in the universe remains constant."[2]

Silliness aside, we ought to be able to say something about how we and our students are doing with all these new programs. Surely, we think that the treatment corresponds more to modern science. Can we not testify that the students are more enthusiastic? In some cases, with or without a contest, is there not increased enrollment? Do not the students spend more time on physics and ask more questions? Try making claims like that, and your questioner will be ready with the perfect squelch: It was only the Hawthorne effect.

The Hawthorne, or halo, effect was named after a famous investigation that started in 1925 at the Hawthorne plant of the Western Electric Company in Chicago. A team of psychologists and what later came to be known as industrial engineers, attempted to correlate output of assembly line components with various factors concerned with worker com-

1. Warren Hipsher, *The Science Teacher* 28, 36–37 (Oct. 1961).
2. National Science Teachers Association, *Theory into Action*, 1964, p. 50.

fort. They went about all this in a very scientific way. There was an experimental group and a control group, and statisticians busily used normalization factors and correlation tests. The workers filled out questionnaires about their health and daily habits. In the first of a long series of tests, the illumination was increased in the work room of the experimental group. Production went up! It went up for the control group, too. Then the experimenters began decreasing the light intensity, step by step. Step by step, production went *up* again, with both groups. For five years the tests continued, getting more and more complicated and producing strange and confusing results. One group of young women workers was given unheard-of privileges—a five-minute break in the morning and again in the afternoon, and even Saturday morning off for a while. The happy girls produced more electrical relays. When conditions were changed back to the normal, respectable 48-hr week with no breaks, the girls still produced more relays.

Management and the research team gradually realized that human beings are very complex creatures. One could even support the hypothesis that being friendly to the workers and giving them a sense of partnership in undertakings would produce more relays, regardless of other conditions. Hence, the term "Hawthorne effect" has come to describe the psychological interactions between human subjects and investigators. A class of students who know that they are part of a new program or a research project will usually respond with interest and excitement. Anything would be better than the usual routine. A teacher who is enthusiastic about a new topic or a new method will find all his expectations fulfilled. The students will also be enthusiastic. Thanks to the Hawthorne effect, all education experiments succeed. Have you ever known one to fail? (What would you say—"Sorry, parents and voters, we sure goofed on that one. Your kids are in trouble but we'll go back to the old system and try to get them up to norms next year.")

The Hawthorne effect is closely allied with the Pygmalion phenomenon. As reported by Robert Rosenthal and Lenore Jacobson (*Pygmalion in the Classroom* [Holt, Rinehart & Winston, 1968]), teachers' expectations concerning their students appear to influence student progress. When teachers were told that certain students were about to bloom intellectually, they bloomed. This happened even in terms of tests, such as IQ, that the teachers themselves did not grade. Indeed, in one extension of this type of study, some graduate students were given two groups of identically bred rats. The students were told, however, that one group had been bred to be maze superior and the other group was maze stupid. Sure enough, that's the way the students found them. The ones that they thought were maze superior did better in learning to run a maze. As further investigation showed, there was apparently a difference in the way the students handled the two groups of rats. (Would you fondle a rat that was maze stupid?)

In a grade school science fair, a little boy had two flats of grass on display. One had a sign saying, "This grass was loved." The other one had a sign saying, "This grass was hated." The magic worked, too. The grass that had been loved was thick, luxuriant, and green. The other grass was sparse and the soil was cracked. "The power of love is very great," remarked the judge, "but did you water each of these equally?" "No," said the boy. "Why should I water the grass I hate?"

The chances are that we will never be able to prove that the innovations in science teaching are "good" or "better." We will always be subject to the criticism that our interpretation is clouded by the Hawthorne effect or the Pygmalion fantasy. But is that so bad? If we can demonstrate the legitimacy of the science involved, if we expect our students to do well, and if they do get interested, isn't that all we can expect? Perhaps we should have innovations every few years just to keep the Hawthorne effect continually in operation. In physics teaching, as in most human endeavors, it's the loving that counts.

November 1969

Clever Hans, the Perceptive Student

Today's cautionary tale is drawn from the annals, not of physics, but psychology. It concerns a horse. Clever Hans, the horse of Mr. von Osten, lived in Berlin 65 years ago. He was called "clever" because he could count, add, subtract, multiply, with whole numbers or with fractions, could pick out colors upon command or people after looking at a photograph, and besides numerous other talents had a discriminating ear for music. Now, in itself, this is not so marvelous. History and folklore are filled with stories of clever animals, including two French dogs in the nineteenth century who could play dominoes. Clever Hans became a celebrity because for some time no one could figure out how he was able to perform his tricks. Eminent psychologists, horse handlers, and other scholars observed and studied the animal. Their preliminary reports asserted that no trickery was involved, and that so far as they could determine, Hans was receiving no signals from Mr. von Osten. It appeared—and this was the crucial issue—that Hans had not been trained, but educated!

In a book published in 1911 and recently reissued (Oskar Pfungst, *Clever Hans* [Holt, Rinehart and Winston, New York, 1965]), the whole story of the horse and his trials is told. Pfungst was a student of Stumpf (we are not making this up) who was an eminent psychologist in Berlin at the turn of the century. Stumpf had been stumped in his own attempt to explain the ability of Hans and passed the problem on to Pfungst. In a classic application of rather primitive scientific method, Pfungst finally found the answer.

Hans communicated with humans by tapping his foot, shaking or nodding his head, or by retrieving and pointing with his mouth. His foot tapping was particularly accurate and impressive. Not only could he do most kinds of ordinary arithmetic, as long as the numbers weren't so large as to tire him, but he could also tap a code to spell out words. Naturally, he could read, though he preferred German script. So did Mr. von Osten. Given a musical chord, Hans could tell whether or not it was harmonic. Indeed, if it were not, Hans would tap out which note was at fault and what its replacement should be.

The presence of Mr. von Osten, contrary to your suspicions, was not really necessary for the performance. Hans obeyed instructions and worked his marvels for many other people, though not all. The human communicator had to stand to the right of Hans and not very far away. Skeptics and men honest beyond challenge proposed questions to Hans and swore that they gave no guiding signal to him. Mr. von Osten himself, and note this carefully, was a retired schoolmaster of dignified mien and blameless reputation. He was convinced that he had taught his horse to think, and was so heartbroken after Pfungst's revelations that he died shortly afterwards, still believing in the cleverness of Hans.

Of course, Hans was clever. And perceptive. Pfungst found out how perceptive a dumb animal can be by running a series of tests that gradually reduced Hans' vision. If Hans could not see his questioner, either because it was dark or because he had been fitted with blinders, he lost his ability completely. Furthermore, even if Hans could see his questioner, he answered correctly only problems whose answers were known to the human. Pfungst finally detected the nearly invisible clues that made the horse stop tapping at the right number, or move his head, or walk in a particular direction. The clues were involuntary movements of the questioner, caused by the expectation of the answer and the release of tension when the right move was made. None of the other questioners was aware that he had made such movements, and many did not believe it even after it had been pointed out to them. Pfungst was able to measure the movements, and even learned to duplicate Hans' tricks by carefully observing a questioner. (Pfungst tapped his finger, not his foot. Try it yourself by having the questioner think of some small number. Tap until you notice a slight relaxing of the questioner's head.)

And what does all this have to do with teaching physics? Do you think that your students study physics or you? How many times in recitation does a student grope his way to the answer you have in mind? How often in private conversation does a student seem to follow your reasoning only to fail the question completely when he is on his own? Why is our own enthusiasm or boredom so faithfully reflected by the class regardless of our actual words?

Unless we have dulled their senses to the point where they no longer care to respond, physics students are at least as perceptive as horses. It's a clever student who knows his own teacher.

September 1969

I n the standard sequence of college courses for a physics major, the senior year is traditionally devoted to modern physics and modern physics laboratory. At different schools and at different times, the content of these courses may range from a phenomenological study of atomic, molecular, and nuclear physics to a very mathematical introduction to quantum mechanics. In any case, the topics dealt with are usually ones unknown one hundred years ago. In most introductory courses there is a similar sequence of topics. If you ask a student to find in his textbook the section on modern physics, he will, of course, bypass the sections on mechanics, heat, and electricity and turn to the sections on atoms and nuclei at the back of the book. We even have a name for the large section of our course that is not "modern" physics; all the rest is "classical." The word "classical" not only has a connotation of ancient, or at least before this century, but also implies that the subject is fully understood and therefore closed. Furthermore, we assume that it is necessary to work through classical physics before we can understand the modern situation. If this is the case, three-fourths of our standard course consists of material that is very important and very useful, but very dead. Any romance or thrill of the unknown, any baffling uncertainties must wait until late spring. During the fall and winter, we can present a wrapped-up world to our students. If we know the history of our profession, and if the students are intrigued by tales of the past, we can point out the quaint notions held by earlier scientists and can trace the rise of truth out of confusion.

Fortunately the real world is not like that. Every branch of physics has a boundary on the frontier of knowledge. There are fads, of course, and at any particular time it may seem that the study of particles or of the solid state is the main problem of modern physics; but then somebody comes along and exploits a topic that no one had been concerned with for some years. A new phenomenon is found, a new question is asked, and suddenly it is apparent that we did not know so much after all.

What could be more routine and less interesting than the description of position as a function of time? Boring though it may be, we have to define vectors and time rates of change in order to make use of Newton's laws. How can there be anything modern about defining position and velocity? But what if you try to measure these quantities in a system moving past you? Is relativity classical physics? In only a few introductory courses would most of the students be capable of exploiting this lead, but surely all students in all introductory courses should hear enough about the subject to make them realize that the simple problem really has very complicated consequences.

What could be more mundane than the definition of mass? To be sure, there is the elementary but often troublesome confusion between units of mass and units of weight. The real question, however, concerns the difference between inertial mass and gravitational mass, and this question is profound and very modern indeed. The equivalence of gravitational and inertial mass is the root of the general theory of relativity. There are still unknown features in this business, and important experiments are now under way or are being planned to explore the mysteries.

The subject of heat at least is classical even though the caloric theory was not abandoned until about a hundred years ago. Now we know that materials possess internal forms of energy manifested in part by temperature and in part by physical state. That's just the point: any model that explains what heat does to material must be based on a model of the microstructure of the material. It is not at all obvious how the heat energy is shared by the various internal motions. It is not at all so obvious as many textbooks make it appear that a solid should expand when heated. It is not so obvious why regularities appear in the specific heat values for various materials. Exploration of these matters leads us directly into atomic and molecular physics, including the latest studies in the solid and liquid states. Properly presented, the thermometer is a tool with which the microstructure can be probed.

Ohm's law has long been classical. Why shouldn't current be proportional to the potential difference in a conductor? But why should it, as a matter of fact? In many systems—vacuum tubes, rectifiers, filaments—current is not proportional to voltage. Indeed, is it not remarkable that the simple law holds for electron flow in metals? Doubling the voltage across a conductor should double the electric field at every point in the conductor, and that should double the force on each charge.

If the force is doubled, should not the acceleration be doubled? Since the current is doubled, however, it must be that the velocity, and not the acceleration of the charge, is doubled. Such an effect could be explained in terms of a drift velocity. Perhaps here is another clue to the microstructure of solids.

There is no need to emphasize that the study of light can be as modern as we choose. Geometrical and physical optics merge into quantum optics in almost every experiment. The easy laboratory availability of sources of coherent light leads to questions of current interest both in science and technology.

Physics is not a subject for students or teachers who are content with certainty and absolute truths. The unknown is never far removed from the simplest of phenomena. Most of the experiments done in our classrooms have their counterparts not only in days gone by but in research laboratories of today. It is all modern physics, and we should make sure that our students get a chance to see this.

May 1968

The Power of the Quantitative

> *"Well, ya know, it's like, ya know, a whole lot."*
> *"Sewer leakage pollutes harbor."*
> *"My love is like a red, red rose."*

We wend our way through life on a very qualitative basis. Civilized society attempts to pin us down to particular times, on a particular date, at a particular location, and with a particular amount of money. But we resist those confining numbers every chance we get.

The distaste for arithmetic frequently starts in grade school. By the time adolescence is supposed to arrive, a person's failure in math is explained by Piagetian labels. The student may never make it from the concrete to the formal operational stage. That means that given $x = y/z$, the student cannot solve for z. It also means that many hand calculators have a special percent key for those who have never learned whether to divide or multiply by a hundred.

Fear of math leads to fear of physics, for everyone knows that physics is the quantitative science. The notion that it is the only quantitative science slanders the others but is true enough for most introductory biology, earth-science, and even chemistry courses. It is possible to get through many of these courses by desperate memorization of names and categories, expressive hand waving, and no calculations at all. Not so with physics. All the important physics concepts are described by formulas. Into those formulas must go appropriate numbers, and the final truth is obtained by solving for the unknown. Can physics be understood without math?

Actually, a quantitative description need not require an algebraic analysis. The functional dependence of one variable on another can be portrayed graphically, and the final magnitudes of variables can be understood by comparison with known objects. In the words of that famous question from "What's My Line," "Is it larger than a bread box?" Such a comparison is the beginning of a quantitative approach to knowledge. Indeed, without comparison of magnitudes with familiar sizes, the numbers lose significance.

A sense of the quantitative is a mark of intellectual maturity. To describe a quantity as "a whole lot" is a cop-out of childhood. The word "pollution" is meaningless except in terms such as parts-per-million and appropriate comparison with quantitative standards. Your love may be like a red, red rose, but that identification would not be of much use if a friend had to pick her out in a crowded train station.

A quantitative description need not be fussy. The appropriate use of precision is another mark of scientific maturity. Since precision usually costs money or time, you should get and report only the amount justified by the problem at hand. For some purposes you may want to assume that there are about 300 people in a hall; for other purposes you may need to count the 289 actually present.

In all our school system, where will students learn to appreciate the power of the quantitative? Certainly not in math classes. The math teachers are concerned more with set theory than with numbers and quantity. Social studies texts are almost devoid of graphs and data. English teachers are too nice even to enforce the use of nice expressions. That leaves gym, shop, and physics.

Lord Kelvin proposes a hard doctrine for us. Surely there is more to life than numbers! How can you quantify the great experiences of humanity—love and sacrifice and creative endeavor? As a matter of fact we put a scale against these experiences all the time. For example, at this season, regardless of our cultural or religious heritage, we all know and can appreciate the precise imagery of these words, "For God so loved the world that He gave His only begotten Son." The magnitude of that love is quantitatively described for us in terms of a standard, if not by numbers. The tradition just wouldn't have been the same if we had been told that God sure loved the world, ya know, like a whole lot.

December 1976

Part IV GOOD CHIDINGS

Editorials are supposed to chide the establishment. Look here! That's wrong! That's silly! From Olympian heights let the thunder bolts fly!

Now, chiding implies a particular kind of soft criticism. A constructive kernel should be buried in the sharp-edged shell. Furthermore, the object of scorn should be a common antagonist. The reader should be on the side of righteousness. The other side should be the impersonal system or some parasitic pseudoprofession or perhaps even the assistant principal. The enemy uses a broad sword; the editorialist an épée.

> For what purpose is the information needed?
> If you don't know, you don't know how to proceed.

> But be ye doers of the word and not hearers only,
> deceiving your own selves.
> —James 1:22

You have to read beneath the lines in this business. Every week I get manuscripts filled with words and phrases fraught with cabalistic meaning. Would you believe, for instance, that in using the *modeling method of teaching that the carefully structured development* and *concept flow* would lead through *Socratic dialogue* to a *rich classroom discourse?* Perhaps it would help if I reminded you that a *model is a primary unit of coherently structured knowledge.* Of course, I am speaking only of the *essence of models* that are at the *heart of modeling theory.* These depend on *student-centered* teaching that is *research-informed,* leading to *interactive engagement methods* and *cooperative learning.*

I used to think that models were those ships in bottles, or those skimpily dressed young women in advertisements. When I reached the age of reason, I realized that a falling golf ball might be a first approximation model of a falling human being. For study and prediction purposes, the model would have some advantages and also some limitations. You could get another model by taking a strobe picture of the falling ball and measuring $y(t)$. Those data could then be plotted or turned into an algebraic equation, each a model of each other and of the falling human. I thought that physics mostly consisted of creating such models, and learning the nature of the constraints. It turns out, apparently, that there is more to be known about the method, and that employing it requires months of training. There's a technique in getting the students to *discourse in a rich way* and you must use a *carefully structured development* that has been *research-informed.*

In case you aren't keeping up these days, I thought it might be helpful if I explain some of the new words. *Inquiry based,* for instance. This means that you don't explain things to the kids. You let them fumble by themselves for awhile, and then, in the words of the old teachers' manuals, you "help them to understand that . . . ." In other words, you tell them because the period is nearly over.

- *Appropriate interactive technologies.* This is like the story of the man who had taught his mule to obey oral commands. First, however, the man had to hit the mule with a stick to get his attention. With students, depending on the hour, the main problem is to keep them awake. Unannounced tests will do this, but the students hate these. Note that we said *appropriate*. A technique that you can use with a class of 25 that you see every day will not work with a lecture class of 250 that you see two times a week.

- *Socratic dialogue.* The epitome of the Socratic method is described in Plato's *Meno*. We ran this account in the March 1994 issue of *The Physics Teacher*, word for (translated) word, figuring that it was its own parody. It's hard to do a parody of a parody, so we'll say no more about the matter here. If you really want to understand the method, better look it up in *TPT* or in the original Greek. Suffice it to say that Socrates tried it with only one slave boy. If you have more than one in your class, better forget it. (For research purposes, of course, it's a useful tool to find out what a particular student understands.)

- *Student centered.* This phrase has many different meanings, but when used as an adjective it means "good" and you don't have to know the details. One manuscript I received contrasts "student centered" with "authoritarian, teacher-dominated, transfer model of instruction." That means "bad."

- *Dramatically improved.* This is what happens to your class after you try any new method of teaching. The dramatically part obviates any need for careful statistics.

- *Active learning, or active engagement.* This is what happens if you avoid inactive engagement. The best technique to keep students active was revealed to me by Carole Escobar long ago. "Don't make them sit for more than 15 minutes at a time, and don't let them out of their seats for more than 15 minutes."

- *Peer coaching.* This is where you take a break and let the kids teach each other in small groups. (Some say three people to a group; others argue for four.) In research this is known as a "double-blind experiment," or "the blind leading the blind."

- *Hands-on experiences.* Don't just show them the bicycle wheel and the rotating stool; let each student feel the phenomena. The bigger the muscles, the better the learning. Keep your own hands off, however.

- *Structured reflection.* This describes the agony a student goes through upon leaving an exam.
- *Discover relationships.* This phrase has nothing to do with genealogy, but describes the way students can discover Newton's laws and String Theory all by themselves.
- *Conceptual understanding.* The student will know that there are atoms and have a feeling that they are very tiny. Obtaining such understanding avoids the *deficiencies of the numerical problem-solving approach,* which was castigated in another manuscript I received.
- *Constructivist strategies.* This method requires the teacher to find out what prior knowledge and misconceptions each student has. The misconceptions must be confronted, presumably to be demolished. Then each student can construct the accepted paradigm for himself or herself, unless you go in for radical constructivism, in which case any paradigm goes, including the original one.

I hope that these explanations clear up a lot of confusion that some of you may have had. Unfortunately, the examples given here were coined just during the last few years. By the time you learn them there will be a whole new set of buzzwords. Take *teaching,* for instance. You probably thought that you knew what that meant!

March 2000

Resortivism—the Library Theory of Learning

t's hard to keep up with the new theories of learning. Only yester-century we were all satisfied with the *Faculty Theory of Learning*. Each person had various faculties of learning—a logical faculty, for instance—and each faculty needed separate training. That certainly seemed reasonable and was a good excuse for teaching mathematics to everyone so that they would learn to be logical. In the early 1900s, Thorndike showed that it wasn't so. A student trained in algebra could not necessarily transfer the skill to legal wrangles or even to geometry.

Once we lost our faculties we turned to *Nature Study, Objectives, Utilitarianism, Piagetian conceptual development, Processes, Mastery, Individualized instruction, Group instruction, Constructivism,* and now the latest thing, *Pieces of Knowledge.* You make your reputation in this field by inventing an exotic or awkward name for a process that every-one already knows. It's the *New* on the package that's new. That's what's new.

I just happen to have my own theory of learning, which is based on *Need to Know.* It is called *Resortivism, the Library Theory of Learning.* In my theory, the brain of a newborn is like the shelves of a library. An astonishing number of books are already in place with a very workable filing and access system. Over the coming years there will be an expan-sion of shelves, stored books, and a continual revamping of the index, categories, and retrieval methods. Some of these changes accumulate until there is a sudden revamping of a whole section of the files, giving rise to the abrupt changes in conceptual ability noted by Joseph Henry, Jean Piaget, and any observant parent. The usefulness of the library depends on the number of books and the speed of access. Individual *Eurekas* occur when a new connection is made between shelves in one's personal library. Genius consists of being able to make connec-tions that did not exist before in anyone's library.

A lot of shelf filling happens automatically, just as it does in our school or home libraries. The books pile up, most of them junk. Fur-thermore, the science books may be mixed up with the Harlequin ro-mances, perhaps with a history text tucked in at random. It takes energy and discipline to keep the shelves orderly. It takes skill and help and constant use to install and upgrade an efficient filing system. Furthermore, in many cases a haphazard arrangement works perfectly well because as long as nobody disturbs the geometry, we remember where we left that lone history book.

A theory of learning demands a theory of teaching. The role of the teacher is partly to add some more books to the individual student library, but mostly to induce a personal revamping, or *resorting,* of the retrieval system. (Hence the name. We could not really call it *revampivism.*) The teacher can present examples of better systems, but in the final analysis it is the student who must rearrange the books and practice using the new index. These processes are painful. How does the teacher persuade the student to go to all the trouble?

There must be a need to know and a need to be able to do. Very young children or gifted older ones create their own need. Most people need a teacher to make suggestions and create the need. At least they must remind us. How? Well, fear sometimes works. Fear of the birch rod, fear of the dunce cap, fear of the failing grade. These are all hallowed by tradition. The overt inducing of fear has fallen out of favor these days, at least in American culture, but of course it is still a common and powerful tool. I don't advocate it, but I do note it.

Another way to create need is to dangle an immediate or ultimate reward. That's the way to get pre-meds to memorize enough physics to do well on their Medical College Admission Tests (MCATs). If you don't pass, you don't go on to med school, and thus kiss that country-club membership goodbye. The principle of ultimate reward is closely akin to that of fear.

How else? Ah, there's the art of the teacher! Some use enthusiasm— it's contagious. Some use drama. Some use the rituals of Socratic questioning. Others use organization, such as individualized, or small groups, or big groups. Some simply state the facts, while others present paradoxes to confound the students' previous systems. It really doesn't make much difference as long as the method entices or bullies the student into practicing the new library retrieval system over and over again.

The difference between the expert problem solver and the novice is that, given a new problem, the expert has solved a hundred very similar ones before. The physics expert has fumbled through all the permutations of questions about Newton's laws as a freshman, sophomore, upperclassman, and then as a teacher's assistant and professor. Propound a problem and the experts rapidly run through their mental card indexes, select the similar solution from the right book, and demonstrate mastery—as long as the problem isn't too different (in which case you must call on a different faculty). Clearly, the best way to learn a subject is to teach it, because then you will get so much experience that solutions to routine problems seem obvious.

The experts have one other enormous advantage—confidence. They have the nerve to rifle through their collection of solutions, knowing that almost every time the right books will be on the right mental shelves. This is known as a shelf-fulfilling prophecy. Poor novices! A scarcity of appropriate books on the shelf, a poor filing system, and no experience in searching and success in finding. Why should we be surprised if they cling tenaciously to the meager resources and illogical filing systems that get them successfully through ordinary life?

Useful learning takes place only if the student does exercise after exercise in the topic—problems and experimental tasks in physics, five-finger exercises in piano. (As a student, Feynman did not do just the assigned problems in his texts; he did *all* the problems in the texts.) Teaching consists of any method that will get the student to do those exercises. During any given year, when time is short, the best the teacher can do is to teach to the test. If the teacher makes the student exercise on plug-in problems and then asks conceptual questions, the poor student will not know how to access that part of the library. And vice versa. The student who has drilled on conceptual puzzles will not have time to learn how to solve the traditional problems.

In this short space you cannot expect me to expound all the ramifications, implications, and virtues of my new theory of learning. Suffice it to say that the main feature lacking is a list of 137 footnotes and citations. If you wish, you can add such a list at random, since no one ever reads them anyway. In the meantime, don't forget the name—*Resortivism*—and don't confuse it with last year's theory which is now subsumed in mine. Remember, it's *New,* and you heard it here first.

November 1993

P anacea was the Greek goddess of healing. She specialized in cure-alls, good for whatever ails ya. Her cult is alive and well. Almost every day our office mail brings a report of a new technique for curing the ails of physics teaching. At the summer meeting in Maryland we heard about more of these.

Many of the innovations consist of administrative techniques for classes—small-group instruction, new methods for monitoring homework, interactive lectures. Usually these methods are based on current learning theories—constructivism, modeling, cycling.

The usual justification for trying a new style of instruction is the discovery that the old style is not very effective. Ever since the Force Concept Inventory was published in *TPT* (March 1992, pp. 141–158), teachers in high schools and colleges have been flagellating themselves by giving the test and bemoaning the results. Most teachers can't believe that after they have explained things so clearly their students are still doing so poorly. College instructors shouldn't have been surprised. In most university calculus-based introductory physics courses, failure rates of 40% are not uncommon, and everyone knows that only the A and B students have the foggiest idea of what's going on.

Since research shows that the standard lecture-recitation-lab course is not effective for most students, we can fall back on several options. We can keep on with our standard methods and abandon any attempt to reach students who don't respond. To assuage our guilt we can claim with considerable justification that the students do not come to us sufficiently prepared and do not work hard enough when they get here. Some years ago a department head at a major university told me that their engineering college expected the physics department to cull the bottom half of the freshmen engineering candidates.

Alternatively, we can try one of the new methods being advocated in our meetings and publications. There is a long tradition of panaceas in education. A century and a half ago, Joseph Henry popularized the use of slate wall boards in classrooms, and think of all the improvements that has produced over the years. One century ago, Edwin Hall created a list of physics experiments required of all high-school students seeking admission to Harvard. Students have been doing experiments in our classes ever since, many of them from that same list. John Dewey spread the word about hands-on learning before it was known as hands-on, and throughout the '20s and '30s children tilled school gardens and set up school shops. Those were the days when it was thought that the new radio networks could bring the words of great lecturers to a waiting nation, thus transforming our provincial classrooms. The promise was fulfilled in the '50s when television brought us science classrooms at sunrise, and Mr. Wizard for the children in the afternoon.

All of these things were good. All education experiments succeed. But except for the use of the blackboard and the omnipresent physics labs, the innovations faded away. (Indeed, many of the new proposals denigrate the blackboards and computerize the traditional labs.) A case in point is the system known as individualized, or self-paced instruction. Educators have been rediscovering this idea since the beginning of this century. There were showcase schools trying the method in the '30s and again in the '60s. At college level in the '60s it was known as the Keller method, and was firmly based on a learning theory known as "mastery." It had its own mystique and jargon, and woe betide any heretic who misapplied the accepted rituals. I organized a version of this system at Stony Brook for a class of 500 before we had heard about Keller. There were no lectures, no formal labs, and no recitations. Just think of all the problems that solved! It was, however, a highly organized system that required a lot of one-on-one student-faculty interaction. It succeeded very well for three years, but then I went off on sabbatical. The system disintegrated. It required too much work on the part of faculty and students. That's the history of most education panaceas. When the instigator leaves or gives up, the system falls apart. That's if all goes well. If it doesn't, then the system turns into a cult, complete with gospel, true believers, and apostates. It will still fall apart, but it takes longer.

One of the problems with education innovations is that it's very hard to demonstrate their efficacy. Students may do better for awhile, but is it because of the intrinsic nature of the method or because the students worked harder or longer on fewer topics? Was the teacher more enthusiastic and perhaps more caring? Is the system transportable—i.e., will it work in other schools? How much retraining of teachers is required and is that realistic? Is anything known about the long-term effects? Are graduates more competent some years later, and how do we test?

Maybe a chief value of an education innovation is that it is new. For a few years both students and teacher are excited about an adventure that stirs up the old habits. It's like tilling a field. Every few years in education, the ground should be turned over and fertilized. That's part of the function of our workshops and meetings—to stir things up.

I went to Panacea once. On the map of Florida, it's a dot on Route 98 just before the causeway where the Ochlockonee flows into the Gulf. Thirty years ago it looked pretty run down, just a few stores and a gas station. They tell me that things have improved since. More gas stations and at least one good restaurant. Like most Panaceas, there are probably a lot of good folk there, but I still think it's a place to fill up the tank and keep going. I wouldn't want to stay there.

October 1996

A Quantitative Misconception

I n various communiqués about proposed school science curricula, we read that physics taught at the seventh- and eighth-grade level should be conceptual, followed by a quantitative treatment of the same material in senior high. There seems to be general agreement that conceptual physics means qualitative and easy, and that quantitative means abstract, mathematical, and, therefore, hard. Even at the college level we hear arguments that students should first study the concepts and then later tackle the grungy mathematical details.

I don't understand this concept. Here at the *TPT* office we continually get manuscripts from people who have just discovered that students have misconceptions, politely called alternative conceptions, about the way the world works. The problem appears to be endemic with old college students and young grade-school children and with every age in between. Researchers keep discovering what good teachers have always known, that our physics concepts are not intuitively obvious, and are not transmitted to our students simply by telling them. Consider that the concept of energy in our modern terms was unknown by Newton, who was not all that stupid. Unfortunately, in spite of all the research in probing student misconceptions, we have received precious little advice about how to warp student notions into our adult paradigms.

Conceptual physics isn't easy. The elementary concepts are profound. For instance, the nature of mass is still a mystery. Consider the complexities of the concepts of energy and mass raised by Ralph Baierlein in the March 1991 issue of *TPT* or by Arnold Arons in the October 1989 issue. Can we fall back on the oft-heard excuse that introductory students need only be taught a first approximation to the "truth"? Beware the moral of Mark Zemansky's quatrain in the September 1970 issue of *TPT*:

> *Teaching thermal physics*
> *Is as easy as a song;*
> *You think you make it simpler*
> *When you make it*
> *Slightly wrong!*

But isn't it good strategy to teach with a cyclic curriculum? Shouldn't we run through the concepts first in a qualitative way, with lots of show and tell? Then, when the children have reached the age of reason, we can give them the full mathematical treatment. According to Eddie Boyes in his article in this issue on American and English physics education, the cycle system that they have been using in England creates problems of its own. He points out that during the earlier treatment, with teachers who are not physics specialists, " . . . there will have been plenty of time for students to be 'turned off' physics because they see it as being too difficult, and they will have developed misconceptions, difficulties, and prejudices against the subject."

Maybe the misconception is in the definition of "quantitative." Quantitative is not the same as mathematical. In fact, most mathematicians are not particularly good at dealing with things quantitatively. Quantitative means knowing (and measuring) the sizes of things, not just in terms of numbers but in terms of the sizes of other, familiar things. Quantitative also involves the way one quantity changes size when some other quantity changes. Relative size and functional dependence. That's what quantitative science involves. In those terms, elementary-school students and their teachers can and should learn quantitatively. By doing so they bypass the Piagetian obstacles of the slow development of concepts. Children can measure and compare quantitatively long before they can understand the simplifying and sophisticated concepts that we have developed. Children can get an operational feel for mass by measuring masses. But don't ask them to fill in the blanks to define mass. In junior high they can measure energy as it transfers from one form to another. But don't ask for a definition of energy on a multiple-choice test. The practice of working quantitatively leaves students with the power to do things. They can manipulate phenomena and when the time comes can construct mature concepts based on personal familiarity with the variables involved. Young people who have only memorized the words of a concept without any experience in handling the magnitudes involved are left powerless.

Of course, there is a shifting boundary between quantitative and mathematical. A functional relationship that seems concrete to one person may appear abstract to someone else. Classicists usually complain that physicists put Descartes before Horace. There is this unjustified fear that quantitative means mathematical. Back in the late 1930s, Einstein spent a summer on the North Fork of Long Island, New York. There he made friends and played violin duets with a local shopkeeper. One day Einstein offered to explain relativity to his friend who protested that he wouldn't understand because he had no math. Einstein answered, "I can explain it to you without math," whereupon he started writing down equations. "No, no," said the shopkeeper. "I really don't understand math." "But this isn't math," said Einstein. "This is just algebra."

So there is a conceptual misunderstanding. From my point of view, after many years of teaching and writing for students of every age group, the simple concepts of physics are tough and subtle. The development and understanding of these concepts requires maturity and years of experience in handling phenomena quantitatively. On the other hand, learning quantitative physics is child's play. And it should be.

November 1991

I n the classic Greek plays it was customary to bring in a god at the end to wind up the story and explain what finally happened to the characters. Since the god arrived suspended by a crane, this method of plot resolution is known as *deus ex machina.* We have a similar problem and a similar solution in education. The whole enterprise is so complicated, and so mixed up with past events, and present problems, and future needs that there seems to be no salvation. Our freshmen are not prepared for the calculus-based physics course, which should therefore be modified in some way. Or perhaps, the fault lies in the high schools that did not adequately prepare the students. In turn, the high schools fault the earlier grades, whose teachers blame society and the parents. Faced with these unsolvable problems, professional educators have turned time after time to novel administrative methods or mechanisms. Failing to find any health in the characters or the plot, they (we) have sought salvation out of the machinery.

(In case you are consulting your Latin dictionary, *salus* in Church Latin means salvation in the spiritual sense as well as health in the original and more literal sense. It turns out that the Romans had no concept of spiritual salvation, and hence no need for such a word. See John 4:22.)

The primary administrative problem in education is, and always has been, how to get the student to work harder with less expenditure of teacher time. At least, the teacher wants to minimize the time; the administration wants to minimize the expenditure. One of the earliest forms of salvation for this purpose was the Lancasterian system of the early nineteenth century. The paid teacher trained a cadre of older monitors (unpaid) who in turn taught the younger students. Graded schools started at about the same time, eventually winning out over the monitorial system. One teacher could instruct forty or more students of the same age more efficiently.

In the 1920s, with the assembly line so successful in industry, efficiency experts turned their attention to the schools. Time studies showed disgraceful inefficiencies in instruction, with the teachers often repeating lessons and the students failing to remember simple conclusions. In response, progressive educators pointed out that many students did not see the relevance of learning to their own lives, so new books and activities were created to bring the school closer to everyday life. Of course, reaction set in immediately. Many parents complained that although the children were learning to garden and hammer, they were not learning to spell. Back to basics. This seesaw of school practice and expectations has been going on ever since there have been schools. Each ten years a previous fad is discovered and renamed. In a curious contradiction, the school system, although accused of not teaching well, has nevertheless been forced to take on the responsibility of teaching health, sex, driving, and other life-saving topics.

You may remember that a few years ago the solution to our problems was to provide the tools for *Mastery Learning*. This was best done with *Individualized Instruction*. For efficiency, however, you may want to try *Interactive Lecturing*. Failing that, resort to *Group Learning*. The current model calls for four in each group, one to read the instructions, one to press the buttons, one to take the data, and one to ask the teacher for help.

Salvation also comes *with* machinery. The all-time classic was and is the chalkboard, first popularized by Joseph Henry 150 years ago. Does there exist a classroom today without at least one chalkboard? If there is a chalkboard, then the rest of the plot is determined as if by fate. In front of the chalkboard there is a teacher standing. In front of the teacher are students sitting. But to save the teacher time, along came radio. Way back in the '20s it was realized that radio would allow lectures to be heard by thousands. Next, it was predicted that salvation would come from the motion picture machine. Lecture demonstrations could now be seen by millions. TV was just a plot refinement, but a potent one. Such a short-lived fad! The interactive computer has now arrived and just in time to save our educational system and turn it into way stations on the information highways of the future. The light at the end of the tunnel is blinding.

I prefer plots where the actions stem from the nature of the characters, their strengths and weaknesses. Life isn't like that, of course. In real life, the gods do come swinging down, saving some and dooming others. However, I don't think they make any difference in the long run. If we were starting all over again in public education, and I knew nothing of the past, I would be fascinated by some of the administrative mechanisms that have been proposed or are now waiting in the wings. But in these latter days I am skeptical of easy solutions. Here is a touchstone with which you can test proposed educational panaceas. If the method causes the student to work harder, and if the method increases student-teacher contact, then it may be worth trying. Otherwise, laugh the apparition offstage and get on with the play.

May 1994

Research Has Shown That—

We don't receive many manuscripts on educational research, but we read a lot about it in the newspapers. Of course, we get and publish lots of notes about some classroom techniques that the author has tried with great success. Most of these make no pretense to being part of a research program. It's more a case of "Hey, look! Here's something that worked well with my class, and you ought to try it too." We welcome reports like this. They're the meat and potatoes of this magazine and of our association meetings.

We do, however, receive quite a few manuscripts that justify their particular approach with a phrase that has a canonical ring to it. It is a phrase that sanctions whatever follows with an aura allowing no dissent. *"Research has shown that . . ."* Research has shown that when people read this phrase, they are supposed to automatically accept the argument as an article of faith. After all, we physics teachers are creatures of science and revere the majesty of scientific research. If someone assures us that "Research has shown that . . . ," who are we to question?

Have you ever thought of the pitfalls of doing educational research? To begin with, you have to deal with human subjects, both with students and with teachers. It might be easier with rats, but we don't use them in introductory physics. Even with rats, you have to be careful. I remember seeing a sign on the cages of a lab at M.I.T. "John, don't pet the rats." It seems that lab assistant John was dealing with two genetically different sets of rats, one maze smart and the other maze stupid. He had been friendlier with the smart rats than with the stupid ones, and it apparently had warped the experimental results. You can understand how that might happen. After all, would you pet a stupid rat?

In dealing with these human students, you face some tricky problems that border on questions of propriety if not morality. Surely you can't do something that may harm the students, but how do you know before you try your new panacea that it won't indeed mess up a class for a whole semester or a year? Remember New Math? That silly venture ruined a whole generation. Mathematica Culpa! (The mathematical association formally admitted their sins at the Euclid Conference, Euclid, Ohio, in 1975. Their excuse was that although the curriculum was good, the propagation system into the schools had failed.) Down in the elementary schools we have the perennial struggle between doctrinaire devotees of *phonics* and *whole language.* The November 17 *New York Times* described another such imbroglio now going on in Westchester, New York. Each side claims that *research has shown that . . .* Needless to say, there has been no valid research on either side.

O.K., you've selected or captured your subjects. How many? Two is not enough. Two hundred is a lot of trouble. In either case, are these typical students? There's the sampling problem—girls, boys, ethnic, socioeconomic background, academic distribution. Will results that you get with 20 have relevance to classes of 200?

Having solved these problems, repeat for your control group. Wait! Who's going to teach your control group? If you are, there's the inverse rat problem. Don't you favor one method over the other? If a colleague is going to teach the control group, how will you normalize the factor of teaching skill?

What sort of test are you going to use? Back in the early 1960s there was great pressure to show that the new PSSC course was either better or worse than traditional high-school physics. In an article in *The Science Teacher* (28, 36 [Oct. 1961]), the authors concluded that their *research had shown that* students taught the traditional course did better on a final exam. Of course, the exam the researchers employed was one designed for the traditional course.

Will your new method take up more time than the standard methods? If you spend the whole fall semester on Newton's laws, your students will probably do better on an exam of Newton's laws than the students in another class who also studied energy, waves, and thermo. About twenty-five years ago I created an impressive demonstration of the success of individualized instruction for a large class in college physics. The mechanism bullied students into doing a lot more work than they usually do, and also required (or lured) the faculty into doing more work. We kept it up for three years but were too exhausted to continue beyond that.

Long ago it took the cooperation of the whole nation and tens of millions of dollars to test the efficacy of the new polio vaccines. Compared with education, medical research is child's play. Can you imagine organizing a double-blind experiment in education? Still, it's a good thing that people are doing experiments and studying the process of learning. Most of the results, such as the conceptual development levels of Piaget, are useful reminders of common folk wisdom (Joseph Henry, 1st Corinthians 13). It's hard to point to any revolutions in educational practice that have come from such research, but in the newspapers we will keep hearing about such future revolutions, over and over and over again. Research has certainly shown that.

January 1994

S chool days are here again! Stretching before us we have a whole new year in which we can try all the wonderful teaching methods, computer systems, and psychological insights that we have been hearing about. Gone are the days when we would lecture to students, pouring information into their gaping ears and writing indelibly on their *tabulae rosae.* Banished are the age-old difficulties of teaching students how to graph or use algebra, all done easily now by computers. Discredited are old-fashioned exams and tests characteristic of false assessment and not based on portfolios and constructivism.

So how is it going to be in your classes this year? One day last spring I had occasion to tour the classroom floor of our physics building. We allow students in math, sociology, and other arcane subjects to use our classrooms. In every room, regardless of subject, I saw a teacher standing at the blackboard, chalk in hand, talking. The students were in their standard positions, sitting, some paying attention, some even taking notes. The only difference from one room to the next seemed to be in the way the chairs were arrayed. The liberal arts students tend to sit in semicircles. In one of the rooms there was one professor, talking to the blackboard, and only one student, who was looking out the door and waved to me. I continued my survey in the philosophy building next door and did see one system of instruction that was different. A group of students were sitting around a table listening to a young man read an essay. It may have been a test in reading.

About 90 years ago my parents attended Mansfield State Normal School in Pennsylvania. The Normal schools were established to set a new norm in teacher education. While there were courses in the very latest theories of pedagogy (the faculty theory of learning was then under attack), the curriculum consisted mainly of subject matter courses. For instance, my mother who was training to teach elementary school had to take a full year's course in physics. (She hated the professor, and thus the course.) If by means of time travel you were to walk into that old classroom, you would feel perfectly at home. The student desk chairs would be arrayed in rows, most of the demonstration apparatus would be familiar, the chalk would be waiting at the board. You could probably use most of your present notes, though you might want to brush up on hydrostatics and certain types of very useful simple machines.

Last year I sat in on some high-school science classes for middle-ability students. It was a disheartening experience. The kids were bored, the teachers were bored, I was bored. Sometimes the teachers would ask questions but it never stirred up much response. These teachers knew very well the advantages of providing hands-on activities, but they also knew the difficulties. There were equipment problems, there were administrative requirements about finishing the syllabus, there was a lack of passion.

It's very hard to escape the standard techniques of teaching. The rooms are constructed so that the students can sit and the teacher can stand. One year I was invited to teach a high-school class all year long. The principal made a public relations event out of it—Brookhaven Lab scientist to teach in local school. He was so proud of the exploit that during the summer he had the physics room refurbished. Instead of the large lab room with wooden tables and chairs that could be easily moved, one half of the room was turned into a miniature lecture hall with raised seats. In the other half the principal had installed tiny fixed lab consoles. There was no place to swing a pendulum, run a race, or stretch a Slinky. The course that I was introducing was the new Physical Science Study Committee (PSSC) course, which demanded flexible space for students and for apparatus. We used the halls, the gym, and the playing fields, but it took extra effort on my part and tolerance on the part of the school administration. As everyone who teaches in public schools knows, you upset the janitors or vice principals and you're in trouble.

Actually, we physics teachers are way ahead in modern teaching methods. We've been doing hands-on teaching for a hundred years. Our students are usually in the upper quartile academically and have elected to take our course. In many schools we are tolerated or even encouraged if we try new techniques. Even if the school is in the dark ages, we can brighten the corner where we are.

And from all sides we hear that things are going to get better. There's a brave new world of education methods, and it's right around the corner. We'll have ways of ascertaining what each student knows and programs to match them where they're at. The individual psycho-analysis will also prescribe authentic assessment techniques, which will be tailored to each student's spectrum of intelligence factors. We will be able to do these things because there will no longer be boxlike classrooms, and certainly no concerns about violence in the halls or poorly prepared students.

In the meantime, September is here. Make sure you have enough student chairs, and check the spare chalk in the blackboard trays. It's back to normal.

September 1994

The Rumpelstiltskin Factor

As I remember the story, there was this king who had a new young wife. For reasons I won't go into here (after all, this is a fable for children), the king thought that his wife could spin straw into gold. One night he put her in a room with a spinning wheel and a pile of straw and told her to get to work. As you know, most young queens cannot ordinarily spin straw into gold, and neither could this one. So she cried. Now here's the good part. Out of nowhere there appeared a strange little fellow who offered to do the trick in return for a small favor. Just a necklace, this time. The next morning the king was delighted and all was well. A week later, the same deal, and again the following week. Each time the elf raised the ante, and the final time he asked for the queen's firstborn child. Since she didn't have any children she happily agreed. But then she got pregnant, presumably by the king, and the elf came back to demand payment. The queen cried harder than ever and so the good little elf said, "One more chance! You can keep the child if in three days you can guess my name." The queen called in the secret service, found out that his name was Rumpelstiltskin, saved the child, and banished the elf.

There are several morals to this story, including the need for elves to get their contracts in writing. What concerns me more is that I've seen an awful lot of straw during the past year with a lot of people solemnly assuring me that they are going to spin it into gold.

Illuminated by a thousand points of light, the United States is going to get the best science- and math-education system in the world. Not overnight, of course, but in nine more years. We will be able to do this with falling state budgets because now we realize the misconceptions (or alternative conceptions) that blind our students and can lead them toward self-enlightenment by employing the constructivist theory of teaching, which is the latest thing. And just in time, because next year there will be a new latest thing in education. (You will excuse my use of the jargon here since it so nicely matches the profundity of the ideas.)

At the elementary-school level, the nation has poured twenty-two million dollars into the Triad projects from the National Science Foundation (NSF). The three legs of each of these curriculum development projects were supposed to consist of developer, commercial distributor, and local school system. With cooperation like that, the product would obviously have the ingenuity of the developer, the real-world practicality of the school, and the commercial backing of the distributor. Most of them didn't turn out this way, of course, but who will ever know? The NSF Education Directorate rarely has external reviews about the results of its giveaways and certainly never makes them public. Still, the inverse miracle did occur, with much fine gold being spun into straw.

At the junior- and senior-high level, the spinning is being done by many disparate groups under the rubric of SS&C (Scope, Sequence, and Coordination). The straw here is piled high and deep. The spinning wheels are being provided by NSF and the Department of Education at the behest of the National Science Teachers Association. With SS&C, all science topics will be taught at each grade level from 7 to 12. Somehow, interesting curricula that satisfy this requirement will be produced for this project. Indeed, committees are already at work, and you know how well committees write. Somehow, 60,000 junior-high science teachers will be trained to teach all science topics. This will be accomplished with workshops. Somehow, these new curricula will be hands-on, arranged for heterogeneous grouping, and relevant to all students (taking into account each one's developmental level). Mind you, this is not just one curriculum, or one method. Rather, the project is to be multi-centered, with each school faculty developing its own system in a way that will enhance the particular local strengths in an atmosphere of team work and collegiality. Such happy faculty teams are already at work in Houston, North Carolina, Puerto Rico, Iowa, and particularly, as you can imagine, in California where the original "100 schools" project has now expanded to 214. Lest you fear that these efforts may be uncoordinated, rejoice to learn that a common assessment system is being planned that will make use of interactive video. Furthermore, if all goes well, the teachers can all talk to each other, and solve each other's problems, by e-mail. (If you think that I'm making this up, read Currents, the SS&C newsletter.)

With all these wheels spinning, I say, "Pile on the straw!" We'll turn it into gold overnight, or at least by 2000, or perhaps by 2061. And if you believe all that, you must also believe in little old elves, named Rumpelstiltskin.

September 1991

A new revelation is sweeping the teaching profession. In practice there are variations in details, but that characteristic is perhaps the most revealing aspect of the movement. For all variations succeed! The common thread, apparently the crucial feature, is that during normal class time the teacher shuts up every so often in order to allow the students to react.

There's a flaw in this system, as you can immediately perceive. If the teacher stops talking, there will be less instruction. Fewer topics may be covered. Countering this defect, the proponents of the new system claim that although the students may be taught less they will nevertheless remember more. Or, at least, they will understand what they do remember.

Several of these instructional systems were described at the January meeting of American Association of Physics Teachers (AAPT) in Orlando. Eric Mazur of Harvard has a technique with large lecture classes to get students to stop and think. (Mazur stops; the students think.) They are given a problem or paradoxical point to talk over with the student next to them. Paul Hewitt has used such a system for years. He gets the kids to talk to each other and arrive at a consensus about some challenge point. Then Hewitt calls on one or more groups for a report. Long ago Ralph Littauer at Cornell created an electronic feedback system for a lecture hall. Each student has a five-button response system. The lecturer can propose a short problem with multiple-choice answer to test student comprehension, or can ask, in general, if students understand a point. The anonymity of the system allows extensive and honest feedback.

The stop-and-think system has a slightly different appearance in small classrooms. In these settings the form could best be identified as *collaborative learning* or *group learning*. This style is becoming very popular in the schools for all subjects. The basic idea is the same: persuade the teacher to present a challenge, stop talking, and let the students try to figure things out for themselves—with guidance, of course.

Quite a few of the trials of these stop-and-think methods have been tested with pretests and posttests, and you will not be surprised to learn that the results are almost incredible. Regardless of the type of test given, the students who experience the new system do better than the control students of some previous year.

You will also not be surprised to learn that I have been using this system for many a year, but without realizing its theoretical basis and without bothering to do controlled tests of its efficacy. In my case, the system was applied in large lecture halls, particularly once in an evening course and another year in a class that met right after lunch. During those lectures when I was not busy writing on the board or erasing, I would occasionally look at the students. Some of them had fallen asleep! When the number of sleepers grew too great, I would dramatically call a halt and decree a two-minute break. Then I would bully the students into standing up, chatting with their neighbors, or doing exercises. When they were all alert and bright eyed again, I would proceed to ladle more truth into them.

These pauses were very popular with the students, and the results were unbelievable. When I link my own experience with the reports of similar systems that are now being tried, I realize that there is a common factor that leads to success. It is the *pause* that counts, not what happens during the pause. I am reminded of the advice given me by a very experienced and successful high-school physics teacher. She said, "Never plan on one student activity for the whole period, particularly if the activity requires sitting still. Get them out of their seats soon after class begins, but then don't make them stand up too long."

Let us merge this evidence with common knowledge from other human activities. It is well known that the average attention span for young TV viewers is about seven minutes. We do not yet know whether this is innate or is due to conditioning. At any rate our young people expect that throughout life there will be a short commercial break every seven minutes. Instead of deploring the system and fighting human nature, *why don't we exploit it?* That's what these new educational techniques are doing, and by their own accounts they succeed beyond belief.

I am therefore proposing a project for higher education to be called P̲ause, A̲ N̲ew A̲lternative C̲lass E̲nergizing A̲ctivity. It will not only capitalize on our new understanding of the learning process but will also solve the financial difficulties of our colleges and universities. Every seven minutes during lectures there will be a short pause for a paid commercial. Naturally, we will carefully screen these ads so that they are politically correct and appropriate to the academic setting. Arrangements could be made for a variety of presentations, including letting the professor be the spokesperson, for a slightly lower fee. The proceeds, which I anticipate would be substantial, could be shared between school and faculty in some equitable way to be decided by union negotiations.

It just might work.

April 1992

Modest Proposals

First, we have the crisis. Our students are not doing as well in science and math tests as students in other countries. Furthermore, they are not doing as well as they used to do here. The teachers of science and math are leaving the profession, and no new ones are being trained. In response to the crisis there are alarms and excursions. Now there are stirrings in the land. Out of Washington come rumors of crash programs and cash programs. Oh boy, oh boy, oh boy. Here we go again.

Would you like to know how I would spend the money? I'm so glad you asked. As a guiding rule I would stay far away from high technology. Twenty years ago I claimed that television in schools was a waste of money and probably harmful. Today I feel the same way about computers. If you want kids to have computer literacy, do it on somebody else's nickel—not the science budget. Actually I don't even know what computer literacy means, and I suspect no one else does either. Perhaps it's something like typewriter literacy.

Wherever science is being taught in grades 1–6, we should make immediate changes. In most cases the best thing to do would be to stop it. With few exceptions we should throw out all grade school science texts, particularly if they are hard bound and have colored pictures. We should rule out any program that requires teachers to learn one new word of learning-theory jargon, or students to learn one new science word. We should ban all attempts to teach any concept (atoms, genes, energy) that the youngsters cannot handle and measure for themselves. Similarly, we should rule out the study of any complex mechanism (rockets, volcanoes, atomic bombs) involving more than two variables. In trying to teach the solar system to these little kids, we keep forgetting that they don't know the way to the next town. Furthermore, their teachers are self-selected not to know any science either, particularly about complex mechanisms.

Let's not waste any more money trying to teach science to elementary-school teachers. We tried that 20 years ago and it was a lot of fun, but it didn't work. There are too many teachers and the turnover is too great. We might try to train them better in college but that would take a cultural revolution. The authors of those pretty grade-school science books teach science methods in the colleges.

What should we do in the elementary schools, other than just stopping science instruction altogether? Here's a very simple and narrow suggestion, but one that has worked and can work again. Establish grade by grade goals of what students should be able to measure and record, frequently with graphs. Tie these goals to the ones that already exist—*de facto*—for arithmetic. Provide meter sticks, thermometers, stop watches, pan balances, string, and sticky tape for every classroom. Prepare short direction sheets to the children, not the teacher, for each exercise. In every case the directions would require them to get out of their seats and measure something—usually by pouring it, or taking it apart.

Forgive me. I exaggerate. When I say *every,* I mean *almost.* There are some splendid third grade teachers out there. There are some precocious sixth graders who can do algebra. I'm not talking about your kids. I'm talking about all the rest of them.

Then there's high-school science. The students are older, and teacher training in science is essential. The solution to our problems here is almost opposite to that in the lower grades. With the much smaller number of teachers involved the nation can afford to train them well and keep them trained. The in-service and summer institutes of the 60s produced a whole generation of better educated science teachers. The institutes also forged and maintained links between colleges and schools. The institutes cost money and they were worth it. They should be revived.

The nation must make teaching careers attractive again, both financially and professionally. If differential salary scales are not practical, then specialties in short supply must be subsidized from outside the system. Let the appropriate agencies do it. In the case of science and math, industries, either on a state or local basis, should institute "chairs" for good teachers. The recipients would receive *appropriate* summer employment, useful for both parties. They would also be the year-round liaison between the schools and the technical staff at the industry. If there can be a Henry Ford Distinguished Professorship at Yale, why shouldn't there be a Henry Ford Distinguished Physics

Teacher at West High? Or better yet, a Milford Boiler Plate Distinguished Physics Teacher at Milford High? For the small cost of a subsidy that would bring the teacher's salary in line with that of industrial peers, the industry would reap more than their money's worth in terms of prestige, public relations, and useful service.

What do we do with the junior high sciences? Most of the teachers in that nether world have very little experience with the physical sciences. Their curricular materials are in shambles. A few schools still teach commercially transformed versions of the curriculum projects of the 60s, but most use texts that emphasize memorization of nonsense symbols and rules, some of them factually wrong. In this wasteland, efforts should be made to develop new curricular material and propagate the few good courses that exist. Summer and in-service institutes should be revived for junior high teachers, with appropriate compensation.

There is one simple reform that school systems could institute at little cost, but few have been able to pull it off. In spite of lip service about the need for coordinated K–12 science programs, they don't exist. Each building principal runs an independent show. If the kids have had no sciences in K–6, they can enter seventh grade science without prejudice or difficulty. They may be better off. What junior high science experiences are required for entrance into *your* high-school physics class? It doesn't make sense to have no coordination between science classes from one grade to the next or from one school unit to the next. All science teachers, from all levels, should be working together as one department. That department should also be concerned with science activities in the elementary schools and advise on setting student performance standards. Come on, everyone. Let's take this business seriously, or cut it all out and save some money.

April 1983

T here's a lot that I admire about the new computer-based instruments. They can measure force at a point without having to stretch springs beyond that point. They can measure temperature with little tiny probes that don't appreciably interfere with the temperature of the material. They can measure the position of an object to within the width of a laser beam. They can stretch time or compress it, doing away with the uncertainties of triggering stop watches with laggard fingers. Best of all, they present the results not as positions on some analog scale but with digits. Lots of digits.

But that's just the recording function. Computers can analyze data too. They can add and multiply in separate columns, and then gather the results all together on a graph, plotting the line of least mean squares and yielding the formula of the line. All that's left for you to do is to photocopy it, staple it to the report, and send it in to *The Physics Teacher*. Thanks to the computer, our experimental work is now free of errors. Free at last.

Can it really be so? Do three significant figures no longer represent uncertainty to somewhere between one part in a hundred and one part in a thousand? Doesn't the math program for least mean squares assume that each data point is a Gaussian collection? Isn't it still very difficult to do 0.1% experiments—even 1% experiments? Most of the research papers I see in technical journals still show error bars on their graphs and if they don't, the researcher has to explain.

Maybe it's a mistake to let beginning students use instruments that yield fine precision and are calibrated to provide great accuracy. There is some point in learning how to deal with errors big enough to drown in. Let students see how 2% uncertainty in Δx and 3% in Δt creates 5% uncertainty in v. Error analysis at the introductory level is very simple and need not involve standard deviations and all the associated arithmetic.

I used to require error bars on all graphs submitted to *The Physics Teacher*. In recent years I've been worn down a bit, but I think that was a mistake. A measurement without error is like a painting without a frame. It might be a great picture, but you wouldn't hang it in your living room. From now on, we want measurements with error analysis and graphs with error bars. Either that or well-phrased explanations. After all, even Mary Poppins wasn't completely perfect.

March 1998

The Teacher-Centered Lecture Method

There are a lot of Powerful Ideas going around these days, all of them invariably leading to Robust Results. One of these is the emphasis on the student-centered curriculum, evidently a hold-over from neo-Deweyism. There are new ways of engaging students in introductory physics, such as group collaboration and microcomputer-based labs. Nevertheless, in most classrooms in the world there are still students seated at desks facing a teacher who is talking. This is particularly true at the college level where there may be several hundred students at a time listening to a formal lecture. Even under these circumstances there are Powerful Ideas to produce Robust Results.

At our meetings and in published articles we have heard about a number of ways to break the nonstop lecture into digestible segments. Some teachers stop lecturing every so often and require their students to answer a question after consultation with their neighbors. The students respond on paper, with flash cards, or electronically. In the April 1992 editorial ("The Pause That Refreshes") I pointed out that it may not make any difference what the students do during the lecture breaks; perhaps the pause simply wakes them up. On the other hand, any system that can bully the student into working seems to be efficacious.

While out of kindness we should extend ourselves to make sure that our students are comfortable and awake, another tactic is to teach them how to attend a lecture. The justification is much the same as we use in teaching them how to read a textbook. When they come to us, they don't know how to read and they don't know how to listen. Reading a scientific text is not like reading Danielle Steele and listening to a scientific lecture is not like watching *Murphy Brown*. To read a scientific text you must have pencil and note pad in hand. You must read what the author claims he's going to do and then see how he does it, reproducing the derivations in your notes and outlining the chapter. Similarly with the lecture, you must take working notes that copy at least what's on the chalkboard. Then within 24 hours you must rewrite the lecture, noting points you don't understand.

The teacher has some obligations with the lecture method. First, there is need at the beginning of the year to teach the note-taking process, perhaps by handing out a sample. Then there is need to enforce student performance by occasional collection and grading of notes at the first or final stage. It's time consuming, but you don't have to do it often. Call it tough love.

There is also the need to give a lecture that's well enough organized and clearly enough presented so that good notes can be taken. The first set of notes on each lecture must be yours. If you use a chalkboard, maintain a ritual of where assignments and notices appear, and a standard typography for topic headings. The writing and diagrams must be visible at the rear of the room. (Use large-diameter "railroad" chalk.) If you use an overhead, make sure the room is bright enough to discourage sleep, and leave time for the students to take notes before you whisk away the transparency. Start out the lecture with a short review to set the stage for telling the students what you're going to cover today. Then show and tell, but never more than one Powerful Idea per lecture. Leave time at the end to tell them what you've told them. Meanwhile, at convenient stopping points, shut up for a minute or two, rousing your students to discuss a throw-out problem or simply to stretch and awaken.

What would a physics lecture be without a good demonstration? Never use video or computer for what you can show directly. However, don't be demonstration happy; there's rarely justification for more than one demonstration per lecture. Whenever possible, use three-dimensional models to accompany two-dimensional chalkboard diagrams. [Otherwise, some student may think that your chalkboard graph of $v(t)$ is an inclined plane.] These models can be made of Tinkertoys, or strings, balls, and flat boards. Under almost all circumstances, make your demonstrations quantitative. That usually doesn't require precise measurement, but the students should get in the habit of seeing physics expressed with numbers and units.

A physics lecture should not be a means of transmitting facts from teacher to student. For that, we have textbooks, which students should learn to study. The lecture should do something that most texts can't—demonstrate an attitude, whip up enthusiasm, show style. You don't have to be a flamboyant showman, but you do have to reveal yourself and your love of the subject. Physicists aren't dealing just with rolling carts and pendulums; we're exploring the nature of the universe! We're in the line of a great tradition and the chase is still on. If we take one step beyond any of the introductory topics we're into current research and mystery. Don't let a class or lecture go by without lifting the veil a bit on what lies beyond. All of science, and physics in particular, is exciting because of all the things we still don't know. When you're facing those hundred students, your powerful idea for the day may not lead to a robust response from any of them. But your personal enthusiasm, if it's sincere, can reach them in a way that no other teaching method can.

October 1995

Razzmatazz and All That Jazz

I was sitting in an American Association of Physics Teachers (AAPT) committee meeting in Toronto, listening to a lot of exciting things that are in store for high-school physics teachers. And I should have been very happy. Every year a hundred science and math teachers are picked for special honors at the White House, and some of our AAPT members have been among the chosen. The American Institute of Physics has a poster campaign underway, urging young people to study physics and maybe even teach it. A large TV organization is planning a series of programs glorifying science, its practitioners, and its teachers. And best of all, the President is going to choose one of us to ride around the earth in a rocket ship.

I kept remembering a cartoon from the '30s. It showed a stage filled with soldiers, sailors, marines, flags, chorus girls parading down the stairs, orchestra rising from the pit, and fireworks going off. In the wings, one cigar-chomping producer is saying to another, "If this don't bring back prosperity, I don't know what will."

Now, don't get me wrong. I love a parade. Some of our classes could use a little of that old razzmatazz. Not many of our schools have a Physics Parents Booster's Club, and maybe the Physics Olympics should have a few cheerleaders. But there has to be a bottom line to the public relations effort. It don't mean a thing if it ain't got that solid learning going on in the classroom. You could put all of us in orbit and it wouldn't do any good if more students didn't learn why it is that things can go into orbit and how to calculate the period. You can outnova Nova with tales of exciting teaching and student discoveries at national laboratories, but back where we work every day there's still $F = ma$. Not even Michael Jackson in a four-color poster is going to jazz more girls into taking physics unless a lot of other changes take place in society.

One of the dangers of hype is that its practitioners come to believe their own words. If we can honor 100 great teachers a year then the teaching situation in our schools must be pretty good. Look how well those winners are doing. Why aren't you doing as well? Maybe next year? Clearly there can't be any crisis any more. This profession doesn't need more money, more people, more apparatus. It needs more medals!

If times get worse, we'll pick out two hundred a year, maybe three hundred. And now that that corner has been turned, and we've given it our best shot, let's start thinking about something else.

In the meantime, are more students signing up for your physics courses? Has your apparatus budget increased? Has the new schedule been adjusted for two lab periods for each class each week? Got your physics institute application all lined up for this summer, subsidized by your school, I suppose? Your salary has increased substantially, I assume. Otherwise we wouldn't have all those new young people training to teach physics. Not here, of course, but they must be around someplace.

And how about your students? Are they learning more than they used to? Are your classes crowded because of all the girls who want to study physics? Do you feel that extra respect the students have for you because one of your colleagues will soon be an astronaut?

Well, well, Madison Avenue wasn't built in a day. These things take time. As I said before, I love a parade, and one of those floats may very well contain bags of money, or new oscilloscopes. Those high-stepping baton twirlers may step right into our classrooms and start learning how to measure acceleration. I wouldn't be at all surprised if that platoon just behind the band consisted of college physics majors marching to sign up for practice teaching.

More music! More flags! Strut your stuff! If that doesn't bring back physics teaching, I don't know what will.

March 1985

Extracurricular Fun and Games

Everybody out for the bus ride! Spring is the time to take a break from the dreary winter routine of studying. It's off we go to the amusement park or the science museum or the nearby research lab. This pause in the day's occupations is particularly popular among elementary school teachers, but is also sought after by science teachers in the upper grades. There is nothing quite equivalent in college-level physics. We have to make do with taking a whole week off and sending our students away by themselves to Fort Lauderdale or some similar educational spa.

Science museums are proliferating. Some of them are independent institutions established as part of a city's cultural ambience. Others are attached to universities or schools, built around some collection or local specialty. At our university we have a museum in the geology department. Along with some rather out-of-place dinosaur bones, it features exhibits of local topography. Thousands of school children stream through every year to study drawings and models of the way glaciers formed Long Island during the recent Ice Age.

Nearby Brookhaven Laboratory has a visitors' bureau that supervises models and demonstrations of nuclear-age processes. They have a Saturday-morning program for junior-high students who come in buses to tour the museum and see a fast-paced show of marvels in the physics auditorium. Liquids change color and gases explode.

A number of our local schools have their own planetariums, one quite professional. Another school is establishing its own science museum, but in this case there is a nuclear power plant in the backyard helping to pay the school taxes.

There's scarcely a community in the United States that doesn't have some sort of science museum. Many of them inherited a collection of dead owls or local minerals, but the best of them are trying hard to mount hands-on exhibits in the mode of the paramount leader—Robert Oppenheimer's Exploratorium in San Francisco. They try to let visitors interact with phenomena, preferably by doing more than pressing a button.

Does any useful learning come out of these exhibits and field trips? Is there any proof, or reasonable conjecture, that a student who visits a museum will do better on the SAT exam? Do more students sign up for high-school physics if they have visited a science museum with their fifth-grade class? For that matter, do students get better grades if they go to an amusement park with their high-school physics class? Are more students lured into taking physics because they have heard about the spring outing at the amusement park? Are science museums and science trips just cultural ornaments of a wealthy society, perhaps a clever form of public relations? (You should see the visitors' exhibit of electric power production run by the Power Authority at Niagara Falls.)

There is something to be said for hoopla and the display of small lightning strokes and real optical images that can't be grasped. As any carny knows, you can't sell candy to the rubes unless you first get them into the tent. But having said that (several times previously), I still think that when the schools make use of science museums in lieu of a day's instruction there ought to be some instruction. Like any other item in the curriculum, the museum visit should be justified in terms of the whole syllabus, should have specific goals, and the results should be tested. The teacher and museum personnel must work together to coordinate the preparation, the visit, and the follow-up. Students should study in advance what they will be seeing and doing in the museum. Most important, each student should go to the museum with one or more real questions to be answered. (Of course, if the questions are trivial, such as the age of the dinosaur bone, the list will be counterproductive. Such lists are fairly common among school museum visitors, but are a travesty of the serious business of learning. They serve the same function as the punch clocks of building guards—proof that you have been at a particular location. Students burdened with such punch lists may find the age that is printed on the dusty caption but fail to see the bone.) There should be planning between teacher and museum about what questions should be asked and how the answers can be obtained. In most cases, only a small number of exhibits should be examined. Finally, unless there is follow-up in the classroom, usually in the form of a test, all is lost.

You may be under the impression that I don't like school bus rides, or roller coasters, or dioramas with stuffed raccoons. Well, in season I can go along with any of these. But I enjoy them more when there is purpose and planning and interesting consequences. Modern-day princes of Serendip discover unexpected rewards along their way only if they are prepared to recognize an unusual event when they see it. With a little bit of planning, fun and games can provide serious instruction and still be fun and games.

May 1991

like T-shirts and catchy slogans. Without a little of the old hype you can't get paying customers into the tent. I'm all in favor of physics for poets, physics fairs, and class trips to the amusement park. Unfortunately, I've just finished grading the first hour exam in our college introductory physics course. I'm grumpy.

The students taking this course are bright. That's true of most of the students anywhere who take physics. The material in this course is demanding but is concerned with exciting phenomena. The course is at the same level that we have been teaching for three decades. The text is loaded with friendly features. In class and lab, the students have handled both exotic and everyday phenomena. They have used both meter sticks and state-of-the-art electronics and have seen how simple atomic models explain a wide range of events. They have also taken part in a carefully planned regimen of lectures, reading, homework, and lab assignments. Their recitation instructors are experienced and make themselves available for help. Their graduate teaching assistants are all American-speaking, cooperative young people who maintain a Help Room. The situation seems ideal for comprehensive learning. For about half the students, it hasn't worked.

Actually, *they* haven't worked. The students who show up at every class, take notes, do their homework, and seek help are doing very well. The students in the other group, who are winging it, are failing. These aren't liberal arts majors who didn't want to take science. All these students got B+ or higher in their high-school science and math courses. They are engineering, or chemistry, or other physical-science majors. They have always done well before by memorizing formulas at the last moment and guessing well on multiple-choice questions. Now the easy method doesn't work for them anymore. It turns out that physics isn't phun. It's hard work.

Accomplishment in any field requires hard work. Music teachers demand and get repeated rehearsals, and practice sessions, and five-finger exercises. Coaches drive their young athletes mercilessly, keeping them for long hours after school and on weekends. McDonald's insists that their employees, young and old, follow exact procedures for every part of their service.

Real science is tough, too. You can't win the Nobel prize, or even the Westinghouse contest, by doing a science project some weekend. Accomplishment in science requires years of patient learning, and step-by-step solving of increasingly difficult problems. That may not be a sufficient condition, but it sure is a necessary one. It's necessary in high school, in college, and throughout all our lives. Ideally, this lesson should become ingrained in all students from kindergarten on. Since we know that doesn't happen, it's up to us, at every level, to deliver the message. If our students have fun but don't learn to do anything, we have failed them.

Of course, while enticing students into the excitement of learning physics, we shouldn't scare them off by making them think that it's too difficult or too alien to their everyday world. Indeed, the everyday world is our main concern, and a working understanding of the physical concepts is within the mental range of a large fraction of the population. Furthermore, there are fascinating discoveries to be made along the way. These may not be new discoveries to the world, in general, but they will be new to the student. In the days of the Physical Science Study Committee (PSSC) physics course, we used to refer to these pleasures as "little Eurekas." The thrill of these personal discoveries is what really makes physics fun. However, let's make sure our students realize the moral of the story of Archimedes and the bathtub. Inspiration came to Archimedes only after a lot of false tries and hard work—probably that's why he was taking a bath! At any rate, it's certain that you can't have even small Eurekas without getting your feet wet.

November 1989

A t the San Antonio meeting there was more discussion about conceptual physics—what it is, if it is, and what it's not. Usually people think that conceptual means the opposite of quantitative, or at least, mathematical. Paul Hewitt popularized the phrase, conceptual, in his book, and yet his book contains formulas. Paul even uses numbers occasionally, although the students don't have to. What is the subtle difference between learning concepts and doing physics?

I am reminded of a story about Einstein and his summer holidays on Long Island. He became friends with a shopkeeper who also played violin and one day volunteered to explain relativity to his new friend. "But I won't be able to understand," said the shopkeeper. "I don't know mathematics." "I can explain it without mathematics," Einstein replied, and then busily started writing down formulas. "No, no," cried the man. "I really have no mathematics." "But this isn't mathematics," Einstein said. "This is merely algebra."

How many "mathematical" concepts are required in order to understand "physics" concepts? Well, it would be hard to get along without the notion of functional dependence. If you change one thing, it changes another. The most primitive such dependence is proportionality. Surely to understand physics at all you must be able to use the fact that for some processes if you double one variable, you double the other. Whether you also have to be able to manipulate $v = kx$ is another matter. Whether you can plot, interpret, and extract information from a graph of $y(x)$ is yet another question. These are three different concepts or skills, and the hierarchy of their development and usefulness is not obvious.

During the second world war the United States Navy taught Ohm's law in three forms: $I = V/R$, $R = V/I$, and $V = IR$. They knew what they were doing, if not why. Two-thirds of a standard human population can't do simple algebra. Nevertheless many graduates of those Navy courses became good electronics technicians and in many cases had a very deep and useful conceptual understanding of electronics.

There's a standard joke in academia. It's the student complaint: "I really understand the physics; I just can't do the problems." We printed this once in *The Physics Teacher* as a centerfold poster. Maybe it is possible to have a partial understanding of Newton's second law without being able to solve for a, given F and m. We certainly observe the converse. We all have known students who can manipulate the formula yet still feel that a continual net force is necessary to keep things moving.

It's hard to imagine a physics course that doesn't bring in the inverse square law. Can that be done without mathematics and problem solving? Now we face Einstein's problem with the shopkeeper. Somehow we have to demonstrate the inverse square law with actual numbers. How does the gravitational attraction change as you move from a distance of 6,400 km to the center of the earth out to a distance of 12,800 km? It may not be mathematics, or even algebra, but it has to be quantitative. Conceptual does not mean *qualitative.* Physics requires measurements and numerical values. These numbers represent concrete operations, or at least they ought to. One of the concepts that a student ought to have about Newton's law is the muscular response to a newton of force. Then there should be sufficient scaling experiences of forces within the human range so that the concept of numerical values of forces below and above the range are comprehensible. Without such conceptual understanding of measured values, based on personal concrete experiences, the mathematical manipulation of formulas is nothing more than a memorized trick. That is all that a lot of students get out of our regular courses. In response to one of a selected group of stimuli questions, they can choose the right formula, calculate the right answer, and bark three times.

A classical example of the difference between the conceptual approach and the mathematical is the relationship between Faraday's work and that of Maxwell. Faraday, with essentially no mathematics, made powerful use of the concept of force fields. Manipulating field lines in his imagination, he could explain and predict a great range of electromagnetic phenomena. Note, however, that he had first-hand, concrete familiarity with the apparatus and phenomena. Note also that his work was quantitative, consisting of careful measurements. Maxwell, the young mathematician, figured out how to describe Faraday's fields with the solutions to different equations. These, in turn, powerfully predicted phenomena that Faraday had not foreseen. Maxwell himself compared the two approaches and gracefully evaluated their relationship and significance:

> Scientific truth should be presented in different forms, and should be regarded as equally scientific whether it appears in the robust form and the vivid colouring of a physical illustration, or in the tenuity and paleness of a symbolical expression.

Of course we want our students to obtain a conceptual understanding of physics. The trap is in thinking that it is easy to do, or that it is a soft approach. Whether the course is called conceptual or mathematical, we dare not divorce the teaching of physics from concrete experiences with phenomena, and those experiences must be quantitative. If we do not provide such experiences, we leave our students with an illusion similar to having listened to great music, whose words and melody we cannot repeat and no longer hear.

March 1984

Too Much, Too Fast, Too Soon

As I write this, we're one-third of the way through the semester and are going great guns. I'm teaching the standard course using one of the standard calculus-based texts. Our situation and experiences are typical of what's happening in most large universities. We've whistled through Coulomb's law and have marveled over the power of the Gaussian approach. The lectures have clearly differentiated between potential and potential energy. E fields and V domains have all been mapped, and now we have polished off both Ampère and Biot–Savart. We're right on schedule.

Unfortunately, the kids are glassy eyed. Except for a precious few, the students have abandoned any intellectual curiosity. Now they are just desperately trying to memorize all the formulas. These didn't do them much good on the exam we just had. If the faculty were not giddy with our race through the text, we would be sobered by the grade distribution on that exam. The majority of the class—and these are the survivors of the first semester—could do fewer than half of the routine problems that we gave them. (It was also frightening to watch 500 students feverishly using their calculators on arithmetic that required nothing more complicated than extracting the square root of 2.)

Was it always like this? My memories of freshman physics at Alma Mater are pleasant. The lecturer was demonstration-happy and never explained much, but it was easy to keep up by reading the text. Practically everybody survived. The few going on to sophomore physics knew that the next year would be rough, but by that time we would be ready for it. I still have the text we used. (In those days students never sold texts that they had labored through.) Although the course was the one designed for physics majors and engineers, it used no calculus. The text made plausibility arguments about functional dependence of variables but then asserted the values of the necessary proportionality constants. Gauss was not mentioned at all with regard to electrostatics. Although Biot and Savart were named, their law was not used for derivations.

How slow and quaint it seems now. Weren't we terribly delayed in our rush for graduate school? Apparently not. Somehow by our senior year we were studying at about the same level as today's seniors. In fact, we had pretty much caught up by the end of our sophomore year. That second-year text and course served as a repeat in depth of special topics from the freshman year. We did not, however, tackle the mathematics of the quantum theory or electromagnetism until we had handled many of the phenomena in laboratory.

Maybe our whole education system is still suffering from the Sputnik syndrome. If *they* got their rocket up before we did, we'd better speed up the training of our scientists. Push mathematical quantum physics down from graduate school to senior year at college. Push relativity and wave mechanics down to sophomore level. In the freshman year analyze all of physics both qualitatively and quantitatively. Why not? We can do it because in high-school physics we have gotten rid of Archimedes' principle and pulleys and have substituted atomic and nuclear physics. The study of simple machines is relegated to the junior high science classes, where, unfortunately, the teachers there bypass such mundane topics in favor of class discussions about nuclear energy.

Many of us have been complaining about the declining abilities of our students in these modern times. Maybe we expect too much. What if we, the teachers, were just starting out to study physics? Could we, transformed into our younger, earlier selves, pass the exams we make up today? Or would we, too, be Aristotelians, with just a surface gloss of modernity gained from turning on the TV? If we had not done well in that first-year physics course, would we have kept studying the subject, or would we now be successful businessmen?

Perhaps the profession should pause and regroup. We have been letting engineering faculties and medical schools tell us that their students must take our fast-paced, calculus-based course. We toss all these students on the same treadmill, even though we know that a large fraction will fall off. Worse yet, we know that many students who stay to the end of the course don't really know what's going on. Maybe first-year physics should contain fewer topics, treated at a lower level. For those who take a second course there could be higher expectations. Certainly in schools where we present too much, too fast, too soon, we lose a lot of our students. Maybe a slow and pleasant start would persuade more students to complete the race. They might even reach the finish line in the same time it takes now.

April 1980

Sleeping with the Fish

Back in the '30s, a young man from the Lower East Side of New York wrote a wildly successful play. When the money came rolling in he bought a boat and nautical attire. He went back to see his mother in the old neighborhood and said, "Look, mama, now I am a Captain." "Johnny," she said, "to you, you is a captain. To me, you is a captain. But to captains, Johnny, you ain't no captain."

Now I am an ichthyologist. Every Wednesday morning I fly up to Boston to consult at one of the world's great aquariums, and every Thursday evening I fly back to my more familiar world. I'm just beginning to know the difference between the sharks and the flounders, whether in or out of the tank. Being a complete outsider, both professionally and geographically, my role is to ask stupid questions about the educational aspects of the aquarium. The New England Aquarium has a sizable education department of very capable people, busily conducting tours and workshops, planning exhibits, and taking materials, animals, and lessons out to the schools.

Throughout the world there is a large and growing movement toward informal science education. Science museums, both large and small, are springing up all over and their nature is changing rapidly. The stuffed owls and static dioramas of yesteryear are being replaced by hands-on exhibits, many inspired by and copied from the Exploratorium in San Francisco.

In almost all these museums there is a continual tension between the conflicting requirements of entertainment and education. Even with endowments and annual fund-raisers, the bottom line must be met by visitors at the admission gate. Besides, it is well known in the carnival business that you can't sell saltwater taffy if you don't get them into the tent. If you make the experience too much like school, your clientele will play hooky.

What's wrong with making museum going an entertainment? No assignments, no sequential topics, no homework, and best of all, no tests. Just fun with unusual phenomena, spectacular happenings, and exotic creatures. Let the lightning flash from the Van de Graaff and let the seal balance a beach ball on its nose. Remember the Princes of

Serendip. Who knows where the random walks through alien fields may lead? Maybe the novelty and excitement will inspire some student visitors and leave them wanting to learn more. Maybe some adult visitor will come away with a better understanding of the world and a firmer resolve to try to save it.

Maybe. But we seldom find out. There is almost no assessment of the outcomes of museum visits. For most school visits, both teachers and students view the experience as a day off, a welcome relief from the usual routine. There is very little preparation and seldom any follow-up. Usually no attempt has been made by the museum or the school to fit the experience into an organized study.

Most museums have special types of group visits, largely as money-making affairs. Birthday parties at the museum are very popular—a little viewing of the exhibits, some arts and crafts, cake and balloons. It may not be educational, but it saves a lot of work at home. At our aquarium, groups can arrange to "sleep with the fish." The kids bring their bed rolls, tour the tanks, make some models, and get fed at night and in the morning. Some real instruction takes place and it helps pay for other educational activities.

Should the museums attempt to do more? Should there be closer collaboration with the schools so that the museum becomes a special laboratory for experiments that cannot be done easily or effectively in classrooms? Without discarding the entertainment value, could museums shift slightly toward the educational techniques of raising questions, providing guided searches, and finally assessing the results?

Museums are ornaments of our civilization. Schools seem to be separate institutions, created only for civilizing our children. But there are science and art lessons in our schools, and our money supports both schools and museums. I have had the opportunity to visit quite a few science museums, and most do try to make themselves useful for the schools. At the New England Aquarium, besides the tours and workshops, we maintain a library and provide kits for teachers. Nevertheless, from my limited experience in this world, it seems to me that the collaboration between schools and museums should be strengthened. Without destroying the entertainment features, there should be a shift toward education. There are many models of how this can be done, one of the best being at the Exploratorium, which lays special emphasis on teacher workshops.

I really have nothing against sleeping with the fish or throwing a birthday party at the local museum. I just think that the relationship between these activities and education is very tenuous. And I ought to know, because now I am an ichthyologist.

April 1996

Reports of innovations in physics teaching methods arrive daily. Such progress! During the summer I attended two meetings where experiments with new techniques were described. Whenever research findings agree with my own prejudices, I conclude that the author is wise and the work well done. If the results contradict my previously held views, I tend to find fault with the experimental methods, if not the good sense of the people involved. In order to make such even-handed judgments I have developed a couple of touchstones to be applied to education experiments.

Touchstones are not used much in physics laboratories anymore. In days of yore, alchemists and jewelers would use a black siliceous stone, much like flint, as a streak plate to test the purity of gold or silver. A streak left by the sample would be compared with streaks left by alloys of known composition. In skilled hands, the method can be quite sensitive for assaying gold. What we need in education are reliable touchstones for assaying experiments, particularly those claiming to be free of dross.

One of my personal touchstones when reading a report on education research concerns the methods of statistical analysis. If test results for two groups are being compared, I like to see distribution graphs rather than statistical criteria such as t-test values. Not only are graphs prettier than cold numbers like 1.43, but they are usually a lot more truthful, at least if they are properly labeled. The sample size, the spread, and the fluctuations are all embarrassingly visible on a graph. We can see immediately that we should not take too seriously a t-test value with three significant figures based on a sample of eleven students.

Actually, my private touchstones make me very skeptical about almost any education experiment where comparative test scores are used as criteria. (Yes, I know that rules out most education experiments.) The literature of such attempts is a wasteland of improperly used statistics and self-delusions. The valid use of statistics is a tough business. The rules and results are useful only so long as the mathematical criteria of randomness and probability are satisfied. In sociological (education) research, the criteria are seldom met. Sample sizes must be large enough to justify use of such parameters as standard deviation. For most applications that means that the distribution must be technically normal. The *t*-test, which William Gosset devised for small samples, is valid only if the sample is selected from the population in a truly random manner. The smaller the sample, the stricter is the requirement for randomness. In education research, samples are often small just because there is built-in prejudice in their selection.

Even if samples are large enough and distributions are normal enough, their comparison is still a chancy business. That's because chance may not be operating in the determination of crucial factors. Did the control group meet at the same hour as the study group? Were both groups equally involved in considering themselves part of a great experiment? How were the experience and enthusiasm of the instructors randomized?

I have one other touchstone that is very nontechnical. If any mechanism of instruction increases student-teacher and student-student contact, it is good; if any teaching technique reduces such contacts, it is bad. Beware the mechanism that requires a large amount of record keeping and paper work. You may not have time to see the student. Beware a method that completely transfers tutoring and testing to clerks or students. You will be lost in administration.

You probably have your own touchstones. You probably won't agree with mine. But we can probably agree that among other similarities between fish and education research, all that glistens is not gold.

November 1977

"**D**o you grade on a curve?" It was one of the first questions that I was asked when I started teaching. Never having taken a course in education, I didn't know what the question meant. Now, twenty years later, I still don't know how to answer.

Apparently there is folklore that the distribution of grades in any class and in any subject fits a predetermined "curve." Presumably, the curve is "normal," which presumably means Gaussian. Then, as every schoolchild knows—but I never learned—the distribution can be divided so that a certain fraction get A's, another fraction get B's, etc. The students at the high point of the curve, and a generous portion on either side, get C.

If that's the system, I ought to know whether or not I "grade on a curve." Well, I do and I don't. I can, indeed, subdivide a Gaussian, but would consider it was undignified to do so for such a purpose. Besides, whether normal or not, my students are seldom Gaussian. In the most recent exam they were almost square. One year I had a camel distribution.

The students don't really care about these statistical quibbles. They want to know if I am going to use a curve, as opposed to "the other way." They have me there, too. As far as I can gather, the other way is, you know, like it used to be in school. Apparently students think that there is some system that yields absolute grades as opposed to relative grades. In New York State this myth is perpetuated by the statewide regents exams, where a grade of 65 is passing. These exams, of course, were not brought down from Mt. Sinai with the other tablets. Instead, they are laboriously hammered out each year by mortals in the state capital. If the exam makers are cunning enough in using their experience and the results of previous exams, they can devise a new one that will not trap too many youngsters below that absolute 65. If they miss, which they do occasionally, and if the howls of the citizenry are loud enough, the absolute grade is adjusted upward with a compensation formula. Any absolute grading system is based on a curve of past experience and informed expectations. If you want to, you can prepare an exam on an absolute grade basis that will flunk everyone in your class or yield them all A's.

About once a month we receive a manuscript that describes yet another elaborate scheme for determining grades. Some of these are so complicated that the student must follow a flow chart to find out whether performance of an extra project is worth the extra credit. To be sure, it's a fact of life that grades can be goads or incentives for learning. I get uneasy, however, when the system gets too mechanical. Rewards for special projects can turn into merit badges. Is there extra credit for coloring the graphs? Furthermore, grade hunger can eat up mechanical grading systems, but that doesn't necessarily lead to digestion of the subject matter. If you have a 1, 2, 3 step system for earning an A, with bonus points for extra work, many B students will be industrious enough to earn an A. But as for knowing physics, they will still be B students.

Keller plan systems of individualized instruction are supposed to produce large numbers of A students by providing opportunities to take tests repeatedly until mastery is attained. If only it could work that way! In the true Keller method, used mainly in psychology courses, mastery is demonstrated by obtaining a score of 85% (or 70%, or 92%—even mastery can be relative) on a multiple choice exam. With such a system you could get a bright chimp to demonstrate mastery after a few tries.

The one feature of the Keller plan that I like is the idea of frequent exams. We should all give exams frequently to let the students know what we consider important. We should grade those exams promptly, and then go over the results thoroughly. Giving an exam only for grading purposes and not also as a teaching tool is a mark of professional inefficiency. There should be opportunity for salvation for a student who has done poorly on an exam. The penance should not be an extra project; it should be the learning of the material missed.

Absolute standards are an illusion of youth. There is no escape from making subjective judgments in grading. No matter what the mechanical system that we use for determining grades—and the simpler the system, the better—we should insist that the resultant number is just a first approximation to the actual grade. At that point, as professionals paid to make judgments, we have to gather together all we know about each student and wrestle with all our experience and prejudices to say: "This is an A student; this one did poorly and should be shocked into action by a C; this one is borderline between C and B, but is trying hard and would be crushed by anything lower than a B.''

But, you say, such subjective judgments may be unfair, and biased, and warped by personal antagonisms or sweet smiles! O.K. Let's use, instead, the strength-test-system that our local phys. ed. teachers tried one year. On an absolutely absolute basis of grading, my son got 82 1/2 in gym. If he had done three extra laps a day, he could have been an A student.

May 1977

Physics Can Be Phun and Also Phony

An ad for another fun-filled physics course came in the mail today. A typical text of this kind deals with subjects that students are really interested in. Based on environmental concerns, such studies are unified by the theme of energy and enlivened with examples from medical school and everyday life. Courses of this type are designed for those who previously hated physics. Naturally, no lab is required. The students are eased into the subject with a core of topics in vector analysis, kinematics, dynamics, conservation laws, the first and second law of thermodynamics, electricity, magnetism, and quantum field theory, all illustrated with cartoons. Then for the second semester there is a choice of more advanced topics. Providentially, no problem solving is required, but the texts usually contain many suggestions for essay-type questions. This is physics without fear, physics without math, and physics that is phony.

Let us backtrack immediately. There are good texts and respectable courses involving every device and method that we have just disparaged. Throughout the country, great teachers have developed clever techniques for enrolling students who normally would not study physics. There are a few books available that are genuinely different, and not simply descriptions of the same old topics coated over with an icing of the current fashion. But they are the exception. Too many courses attempt too much coverage, at the expense of being intellectually shallow. The addition of a theme does not necessarily unify the subject for the beginner. Nor does leaving out math reduce the conceptual difficulties involved in physics. Physics is concerned with the functional dependence of one variable on another. It is a fallacy to think that a paragraph of prose can clear up mathematical illiteracy. If a student does not understand linear relationships in math symbols or graphical forms, then the student almost certainly will not understand a verbal description. If it is true, as the Piagetans keep telling us, that only about one-third of high-school seniors or college freshmen are in the formal operational stage of development, then our standard physics course is inappropriate for two-thirds of our school population. No magic wand will make it otherwise.

Must we conclude that science and modern technology are hopelessly beyond two-thirds of our population? No, there are many useful and profound things to be learned at the concrete operational stage of the Piagetian levels. The prime lesson that the physical sciences can teach at this level is the quantitative approach. Phenomena can be examined directly and the parameters measured. Functional dependence of two variables can be graphed. The order of magnitude of quantities can be determined and used in simple calculations. Probably the most important lesson that we can teach concerns the idea that although we can deal simply with simple models, real phenomena and events are usually very complex. The laboratory, not the text, is the appropriate teaching tool for this concept. For such a lesson to be learned, however, we must exploit, not excuse, the fact that many class experiments do not "work."

We all know of cases where a wide-wonderful-world physics course attracts a large audience of enthusiastic students and the admiration of appreciative principals or deans. Usually in these cases the student can get a good grade with very little work. Usually, also, the students are well entertained and think that at last they understand science. Such courses border on intellectual fraud, and they may also reduce the number of students taking physics courses that are more demanding. A Gresham's law applies here, as well as in economics: easy courses drive out hard courses. Any teacher or department offering a milk-toast course should periodically review the situation. Are the abilities of the students being appropriately used and stretched? Does the existence of the easy course seriously reduce the number of students who should be taking a more demanding one?

The Pied Piper led the children of Hamelin town to believe that with his music they could leave their lowland studies and climb mountains. Do you not think that when the music stopped, the children cursed the Piper for leaving them powerless in an alien land?

April 1976

The Great American New Math Swindle

T he "New Math" curriculum has been a tragic mistake. Never before has a fad in education been able to sweep so rapidly through the schools, altering texts, teacher training, and parental attitudes in the course of one decade. Unfortunately, the changes were propagated by a small group of educators who did not have adequate direction or backing from the professions that actually create and use mathematics. Nor did they have any experimental grounds on which to base their theories of learning.

The deceptively logical basis of the School Mathematics Study Group (SMSG), the Illinois project, and similar programs was that meaningful learning is more effective than rote learning. That seems logical though there is only contradictory evidence for the validity of the notion when applied to actual arithmetic learning. Nevertheless, in the name of this precept all sorts of new and strange nonsense games were wished upon children. In order to understand the true significance of numbers and addition, six-year-olds were taught to mouth some of the words of set and group theory. In order to understand the decimal system, fifth graders were confused by operations with other bases. In order to do simple algebra, fifteen-year-olds were subjected to language and abstractions that baffle college math majors.

Most parents were bullied into thinking that their children were now being taught powerful new methods for this modern age. Parents with technical training in math knew that these new school games were unrelated to real math use. But how could any parent challenge the system? Surely these educators must know what they are doing!

They didn't. In the final report of the SMSG Panel on Research (Newsletter No. 39, August 1972), there is a summary of the research attempts to find differences between the results of the new math and the old. As you can imagine, most of the research was meaningless—not enough students, too many variables, etc. The SMSG panel concludes the same thing: "In most of the studies the number of subjects is too small to encourage generalizations. Almost never is it possible to compare two of these studies, since they use different treatments, different instruments, and different kinds of subjects." The panel also cautions that: "We have nothing to say about problem solving, despite the fact that probably our most important overall goal in mathematics' education is to develop in each student the ability to apply the mathematics he has learned to real world problems. A recent review by Kilpatrick (1969) makes it clear that our empirical information about problem solving is so scarce and disconnected that attempts to theorize about it now would be futile."

We physics teachers need not be without sin in order to complain about this situation because the first stone has already been cast by other people. We might, however, worry about our own glass house. Didn't we make the same arbitrary changes in curriculum starting fifteen years ago? Yes, but there were crucial differences. The physics curriculum revisers represented a wide range of creators and users of physics. The new programs provide information about current technological problems; the old courses were badly outdated in this regard. A practicing scientist or engineer could recognize the material in the new physics courses and even help his children do their homework. Where we erred (or at least where we didn't go far enough originally) was in concerning ourselves only with the upper level students who already took physics. That has changed now, and many physicists and physics teachers have played leading roles in the development of science courses for younger students and for all academic levels.

As far as the high-school physics course is concerned, we were in a much safer position to tinker than were the math educators. If we make physics too abstract or too dull, a few students won't study the subject. Too bad, but at least we haven't affected the whole school system. All students, however, study math, and at every grade level. Mess up one of the three R's, and you have done major damage to the educational system. That's just what happened.

In many colleges there is a current swing back toward linking math studies with applications. This trend should be encouraged at every level and physics teachers can lead the way. Our training provides a natural bridge between the abstractions of a logical, symbolic system and the realities of computations and quantitative analysis. At our own school we have been asked by the instructor of the introductory calculus course to cooperate in the design of applied math materials. If you have ever met a student who could solve $x = yz$ for y but could not solve $F = ma$ for m, why not proposition your local friendly math department to consider your mutual interests? It may be too late for the kids who never learned set theory in grade school, but at least you might be able to teach the physics students how to multiply and divide, make graphs, and maybe even do some simple algebra. To be understood, math must be old and used.

March 1973

The Mystery of Mastery

For years we have been developing and advocating individualized instruction. Now we are told that there is a theoretical basis for the superiority of the method. It is called the Mastery Theory of Learning. The system is explained and canonized in a book by Block, *Mastery Learning* (Holt, Rinehart & Winston, New York, 1971).

Apparently the term "mastery" was first used by Benjamin Bloom, a psychologist from Chicago previously known for his work in taxonomy. The theory is that any body of knowledge can be organized in a hierarchy of learning objectives: A → B → C etc. In a traditional class, the instructor goes marching through the material in a set time interval (one semester, for example) while the students tag along with varying degrees of comprehension. At the end of the semester student success is ranked, and grades invariably form a normal distribution pattern. Instead of this system, says Bloom, why not let each student master A at his own speed, then move on to B, etc. Eventually, each and every student will master the whole subject. There will still be a normal curve representing the results, but now the curve will show the distribution of times needed to complete all the work. In a crude sense we would have a mathematical transform of variables from "fraction completed in given time" to "time taken for complete mastery."

Naturally, a theory like this needs testing. Naturally, the testing has been done with methods based on the theory. Naturally, the experiments were successful. The usual procedure for testing this method (known as the Keller method in its most common form) is to choose an experimental student group and a control group. The control group gets the standard treatment of lectures, recitation, and two or three exams per semester. For the experimental group the course is broken down into ten or more subunits with a proficiency test attached to each one. The students work their way through these, achieving Mastery on each subunit as demonstrated by getting 80% or 90% on the proficiency test. (How I envy these students. After 30 years of teaching Newton's laws I still feel a lack of Mastery.) At the end of the semester both groups take the same final exam. In most cases (though not all) the experimental group does slightly better statistically on this exam.

Well, of course the experimental group does better! They have achieved Mastery, all right, but what they have Mastered is how to take the instructor's exams. This system is known by a slightly different name in New York State. It's called "Reviewing for the Regents' Exam." Any New York teacher worth his tenure knows that not long after Easter he had better start the class practicing on the Regents' exams of the last ten years. By the time the students have achieved Mastery on those they are ready to really hit their own Regents' exam.

In the words of the poet, "Research is Long and Life is Short, Particularly in Educational Research." There are valid reasons for individualized instruction using self-paced modules, but it is unlikely that any profound theory of learning is going to be substantiated by trivial trials of this method. After all, it took millions of dollars and the cooperation of the whole nation to demonstrate the efficacy of the polio vaccine. That's a single, straightforward question compared with the complex problem of school learning.

Still, I must admire the choice of words for the theory and method. Mastery! What a great fad word. What taxpayer wouldn't want his children to achieve Mastery in school? Actually, that's the way I teach and I'm sure that you all do too. And after my students achieve Mastery they're going to work, step by step, on Virtue.

December 1972

was startled the other day to receive a manuscript describing a new teaching method that is being carried out in a "constructivist atmosphere." It suddenly occurred to me that I had not heard anything about constructivism for a long time. I always had trouble with that word. When it was becoming popular, I couldn't figure out what the methodology was or why it was unique. Furthermore, if I tried to explain the term to anyone not washed in the blood, as we Baptists like to say, I couldn't even think of the *word*. I kept coming up with "conceptualism" or "destructivism."

There was a similar problem described in the Old Testament (Judges 12:5–6). The Gileadites were slaughtering the Ephraimites, as was the custom, and the question arose as to how to tell your tribal brethren from the wicked enemy, since all looked very much alike. The solution was to require each suspect to say the word "shibboleth." The good guys could properly pronounce the "sh" sound; the villains could not, and were killed.

For about a decade you couldn't get funding for education research without using the word "constructivism" correctly in the prologue of the grant application. In an earlier epoch, the required words were "mastery" and "individualized instruction" or the "Keller plan." More recently we relied on "group" or "peer" instruction, with the serious debates being whether the group should number three or four, and who should form the group. "Group" is still a potent word, but the current shibboleth is "interaction." The students must interact with the text, and the lab apparatus must interact with the student, preferably in real time. Under all circumstances, the lectures, if there must be lectures, should be interactive. This means that both students and lecturer must be kept alert, or at least awake.

There is a variety of evidence, and claims of evidence, that each of

1. D. Hestenes, M. Wells, G. Swackhamer, "Force concept inventory," *Phys. Teach.* 30, 141 (March 1992).
2. R. Hake, "Interactive-engagement versus traditional methods: A six-thousand-student survey of mechanics test data for introductory physics courses," *Am. J. Phys.* 66, 64 (January 1998).

the latest fads produces superior learning and happier students. In particular, students who interact with apparatus or lecture do better on the Force Concept Inventory exam.[1] The evidence of Richard Hake's meta-statistical study[2] is so dramatic that the only surprising result is that many schools and colleges are still teaching in old-fashioned ways. Perhaps the interaction technique reduces coverage of topics, or perhaps the method requires new teaching skills that many teachers find awkward. At any rate, the new methodology is not sweeping the nation.

No matter what the name, a teaching method has to lure the student to work. For some few students, a good text will do the job. For others, an inspiring lecture may trigger active learning. Some students work better alone, some in groups. In all cases, the student will learn only if there is a need to know. That need is best aroused by curiosity. Failing that, an imminent exam is frequently useful.

Interactive teaching emphasizes immediate response and work by the student, whereas an exam leads to bursts of work and cramming. Most of the interaction techniques require continuous student effort. They also require continuous teacher effort. And there's the rub. Most classrooms, particularly at the college level, are not set up for interactive learning. The geometry and furniture are designed for lectures, with the students sitting in rows and the teacher standing by the blackboard. It takes skill and the right personality to defeat the traditional system.

On the other hand, we all know cases where teachers have exploited the standard system and managed to inspire their students to work hard and learn. It's usually a matter of the teacher taking genuine interest in each student and remaining personally fascinated by the subject matter.

I'm afraid that constructivism is fading fast. It embodied a truism that the teacher cannot "learn" the student. It is the student who must learn the subject and construct a valid mental model. The radical constructivists claimed that one model was as good as another, but very few learned to say that shibboleth.

One of the pleasures of old age is looking back at the rise and fall of teaching fads. The period for each is about 10 years. I can hardly wait to see what the next innovation will be. I sure hope I can learn how to pronounce its name correctly.

September 1999

Reprise (Critical Thinking and Similar Nostrums)

Here we are at the New Year again. Not for us, perhaps, but for the calendar. For teachers the year begins in the fall, and now we're on the homestretch. Electricity and magnetism are practically at our throats. While charging up my capacitors and spirits over the holiday, it occurred to me to list the great curriculum reforms of the last few years. One of our faithful readers had questioned my earlier assertion that every such reform had its predecessor.

How about *Critical Thinking?* Surely that should be part of any modern curriculum, particularly in science. The president, himself, has specified critical thinking as one of the goals of education. I would be in favor of it, too, if only I knew what it meant. Presumably, we do not mean *critical* in the pejorative sense. We do not want a whole classroom of skeptics and nay-sayers, criticizing every move we make. Does a critical thinker take everything the teacher says with a grain of salt? That might be healthy, though we would never finish the work of the semester. Perhaps, by *critical thinker* we mean *logical thinker,* somebody who can put 2 and 2 together, and get an answer somewhere between -4 and $+4$, depending on whether the 2's are scalars or vectors. Surely, therefore, physics is an ideal course for breeding such skills. Would not a critical thinker in physics also be a critical thinker in legal matters? If so, let us train the critical faculties of our students. Unfortunately, this sounds to me like the *Faculty Theory of Learning,* which was pretty well trounced by Thorndike about 90 years ago. (By faculties, we do not mean the problems caused when the dean is fired and loses his faculties.) It turns out that logical skill in one subject does not guarantee logical skill in any other subject.

Let's hear it for *Hands-On Learning.* That's what John Dewey was advocating back in the 1920s. Kids planted gardens and ran their own school stores and were graded on their ability to do things. It's hard to say whether these were *competency-based* tests or *authentic* tests. It is sure, however, that the thesis soon produced its antithesis. By the mid-'20s there was a movement, led by Bagley, called *Essentialism.* It was very much like our more recent reform, called *Back to Basics.* Reformers get so mixed up. Dewey's *Progressivism* of the '30s was transformed in the '90s to *Conceptual Learning.*

Isn't it about time that we had *standards* for education? Every village, county, state, and the great nation itself is spending time and money drawing up standards for our students. And about time! It was only 100 years ago that The Committee of Ten under the leadership of President Elliott of Harvard drew up such standards for the schools. In the meantime, for anyone interested, we have had the New York State *Regents Exams*, as solid and detailed as standards will ever get.

Do you remember *Constructivism?* It seems to be fading now, perhaps in favor of *Bits and Pieces of Knowledge.* (Would I kid you?) Constructivism, at least in its less offensive form, pointed out that each student must construct knowledge for himself. The teacher's role is to learn where each student is coming from and to prepare the right next question at the right time. As Rousseau said in 1762, "The first thing is to study your pupils more, for it is very certain that you do not know them." And as Pestalozzi said in 1775, "A person is much more truly educated through that which he does than through that which he learns secondhand."

The current trend is for *Peer-Group Instruction.* When the blind lead the blind, they may stumble more but they do it with empathy. There was a somewhat similar attempt in the early 1800s, called *Lancasterianism.* A master taught a cadre of older students who in turn led the learning of a group of younger students. It seemed to meet the goal of education administrators—to teach more students with fewer teachers. However, the system turned out to be less efficient than the graded classroom, which has been with us ever since. Of course, there was an antithesis, reappearing every other generation. It was *Individualized Instruction,* known in one of its recent incarnations as the *Keller Plan.* The philosophical underpinning, speaking loosely, was known as *Mastery Learning.* Particularly in physics, where the topics are so sequential, it would seem to make sense not to start dynamics until you have mastered kinematics. Of course, it's hard to say when someone has finally mastered Newton's laws, and the system bogged down in arguments as to whether mastery meant 80% or 90% on some multiple-choice exam. The heyday for individualized instruction was in the 1930s, with the best known example in Winnetka, Illinois. Like most education experiments, it succeeded brilliantly, but failed to propagate.

At the turn of the year, it's therapeutic to look back and to look ahead. Knowing the past should not quench our enthusiasm for the reforms of the future. However, such knowledge may steady our nerves. As Santayana said, "Those who cannot remember the past are condemned to repeat it." Besides, it's always useful to know that a *new model research-based trenching mechanism* is, after all, just a *spade*.

January 1998

Quod Erat Demonstrandum

As every schoolchild knows, Q.E.D. stands for quantum electro-dynamics. In the classical age, however, before the New Math, students of Euclid's geometry learned that Q.E.D. at the end of a proof signified "Quod erat demonstrandum." Note the mood and tense of the verb—"which *was to be* demonstrated." The conclusion of a Euclidean theorem was preordained. This also seems to be true with many experiments in education.

We hear about education experiments all the time. Never a month goes by but we receive a manuscript describing a new organization of topics, or a new way of grading homework, or a new way to replace lectures with individualized instruction. The experiments are invariably successful. Usually the author is convinced that he has found a panacea for our teaching ills. We've tried a few panaceas ourselves, and probably so have you. In general, we're all in favor of teachers continually reviewing and renewing their techniques and trying new things. The problem is, how do you determine whether the new material or the new technique is better than the old? A tempting but misguided answer seems to be that you must test the innovation scientifically. Choose a sample and a control group from your population, present the panacea, and then administer a test instrument (give an exam). The data must then be analyzed with statistical procedures—chi square, *t*-test, or even more arcane devices.

It can't be done. It never has been done. The conclusions of thousands of Ed.D. theses present embarrassing testimony to the impossibility of performing a useful experiment in education, *if* the analysis must depend on honest statistics.

Consider the problems:

1. The test sample and control must be representative of the population *and* randomly chosen. These two criteria are not the same. Random selection will yield a representative sample only if the sample is large. Furthermore, the crucial characteristics of both sample and population must be known in order to determine whether the criterion is satisfied. Frequently, it is not even obvious which characteristics are crucial.

2. The sample and control sizes must be large enough to provide result differences that are statistically significant. A chi square comparison begins to be valid only if at least five points or categories are chosen and only if there are at least five members in each category. Even then, the test presumes that variation in results caused by different sample choices is determined by chance. With human subjects this proviso is rarely satisfied.

3. The most serious obstacle to statistical analysis of education research is the need to control variables. Usually only one variable can or should be tested, but the results are determined by many other factors that either cannot be controlled or are not even known.

Suppose you want to test the differences between lecturing to one group and giving individualized instruction to another. *Who chooses the students?* Do you deny them choice (even of transfer or of dropping out) or do you prejudice your experiment from the beginning?

Who lectures? If you are the lecturer, will you give your usual polished, enthusiastic, carefully planned lectures, with pedagogically clever demonstrations? Will your results be relevant to some other lecture situation where the lecturer is better or worse than you, where the class size is larger or smaller, where the students are younger or older than yours?

Who administers the individualized system? If you are a recent convert, do you personally and enthusiastically deal with all the students, or do you, with scientific impartiality, assign the task to an uncommitted colleague? Will his results be relevant to some other situation where the competency tests were written by somebody else and are available at more convenient hours or less, with more students per teacher or fewer, with other similar courses in the school or as a novelty?

How will you test your results? Student opinion questionnaires can be used instead of accepting apples but no physics teacher with any sense of propriety would try to analyze such things statistically. (After all, we are heirs of a great tradition of using statistics with perception.) The gobbledegook concerning the affective domain is best left to people who have nothing better to do or teach. Learning how students feel by probing them with multiple choice questions has the same relationship to the skill of teaching as painting by numbers has to the skill of an

artist. In both cases, there is an art involved which cannot be characterized by a t-test.

Will you test your experimental results with physics exams? Will the
tests be prepared and judged with double blind methods? Will each
group get the same number of exams, with the same questions, given at
the same time? Has each group had the same emphasis on each topic? Is
frequent testing perhaps a peculiar and important characteristic of
individualized instruction and difficult to administer in lecture mode?
Would not a comprehensive final (bearing due weight in grading) disturb the subtle format of the frequent competency exams? Remember
the classic silliness seventeen years ago (*The Science Teacher*, 28 (No. 6)
36, Oct. 1961) where an elaborately controlled experiment to compare
Physical Science Study Committee (PSSC) and the traditional course
used as the final test instrument an exam designed for the traditional
course which was being taught to the control group! The experimenters
used about 100 students in each group and reported the adjusted means
to five significant figures.

Twenty years ago it took the cooperation of the whole country and
tens of millions of dollars to determine whether or not the polio vaccine
worked. Doing that experiment with its relatively simple and single
criterion was child's play compared with the difficulty of analyzing the
effects of educational methods. Our work is subject to too many variables and too many factors unknown to us or outside our control. Our
ability to try major experiments is limited by cost and by the difficulties
of using humans as subjects.

Can we never try some new teaching method and learn how it
works? We all know we can! The skills and techniques required are the
same as we use in our teaching. We listen to our students and try to
understand them. A good teacher gets lots of feedback. If you are enthusiastic about a new topic or a new explanation or a new technique, try it
and feel the response. If you think the innovation worked for you,
you're probably right. Hang on to it until you get bored and try something new. Remember that all education experiments are successful if
the investigator is enthusiastic. Just don't try to analyze your results
with statistics that apply to dice rather than to students.

Education experiments will continue, of course, with evermore elaborate statistical analysis. A new fad word has recently been coined to describe the extraction of hard conclusions from the soft data of many different experiments. The word is *metastatistics.* Presumably it bears the same relationship to statistics as metaphysics does to physics. With such a tool, conclusions will be unassailable because no one will be able to understand how they were obtained. As an investigator, you will then be able to perform the perfect experiment. First, decide how you want your results to turn out; next, do the experiment (this step can be left out); then apply statistics or metastatistics to the data. Your desired result will be assured at the 0.05 confidence level. Q.E.D.

March 1978

New Fad, Old Fraud

E ducation has a new fad word—competency. It is closely akin to "mastery" which students are supposed to attain, but competency is an attribute that *teachers* must demonstrate. Ideally, teachers would demonstrate competency by having their students attain mastery. Of course, no one knows how to do that and it would probably be harmful if we did. Nevertheless, many states are busily planning to license teachers in terms of competency based certification. (CBC known last year as PBC: performance-based certification).[1]

Strange as it may seem, I was one of the first to administer competency tests (CT). Years ago, through skillful political bargaining, I was elected captain of my sixth-grade baseball team. The chief function of the captain was to choose the pitcher. To be fair, I set up a CT, which in this case consisted of ten throws toward a strike zone chalked on the school wall. Every boy in the class tested his competency in this eminently rational way. The worst possible pitcher won. Even though during the test he pitched the largest number of balls into the strike zone, we all knew he was the worst choice, and he was. That early failure of CBP (competency based pitching) made me cynical about all school tests as well as many other education programs.

But surely we can tell the difference between a good teacher and a bad one! I used to be able to, but that was fifteen years ago when I first got into this business. Take that first in-service institute where I delivered solid and polished lectures about the subject matter in the new Physical Science Study Committee (PSSC) course. Can you imagine, there was only one man out of the fifty high-school physics teachers who knew enough calculus to follow my derivations! He quit teaching the following year—couldn't control his students. At the other end of the ability spectrum was the character who always came late, sat in the front row, and asked questions. Some of them were so elementary I didn't even know the answers. It took me quite awhile to realize that he

1. If you want a less biased description of this movement, write to American Association of Colleges for Teacher Education (AACTE). They have produced a variety of studies on CBTE (competency based teacher education). The address is: AACTE, 1 Dupont Circle, N.W., Washington, DC 20036.

was asking questions everyone else was too embarrassed to ask. In the years since I have often visited that teacher's classes and have also talked with men who used to be his students. I'm sure that he's a master teacher, but I'm not sure that he would do very well on a written CT.

Note, however, that I do claim that this man is a good teacher. Apparently, teaching skill *can* be rated. The trouble is, this man's success depends on his style, which is very special, impossible to copy, and probably not effective with certain types of students. This man is flamboyant and is always "center-stage." Another local teacher who is also very successful with many students is quiet, almost to the point of being shy. Which personality type should we certify for competence? What sort of test can we administer to judge effective personality? Let's face it: teaching is an art, and judgment of good teaching is also an art.

If this fad of competency tests were just another silliness propagated by theorists in the ivory towers, I could enjoy the joke as well as the next man. After all, the cult of efficiency ratings rose and fell in the 1920s leaving the schools untouched. In the normal quagmire of the school systems, competency based certification would also sink without a trace. Unfortunately, there's a lot of money behind this move, because various legislators are demanding new tokens of accountability. The New York State Education Department has mandated that within a few years all prospective teachers will be certified in terms of competency tests. Nobody knows yet who will test for what. Committees are forming, conferences are being held, and new words are being coined every day. It looks as if there will be a whole new industry to draw up these civil-service type standards.

My own suggestion is that the way to make a competency test for teaching any subject at any grade level is to chalk up a rectangle on the school wall and give each candidate ten free throws. It's a lousy way to select a pitcher, but there is absolutely no education research to say that it's not as effective as any other formal test for picking a good teacher.

May 1974

The Order of Nature

The different faculties and powers of the mind are not simultaneously but successively developed, with some periods of life better suited to particular acquirements than others. We ought to follow the order of nature, and to adapt the instruments to the age and mental stature of the pupil. Memory, imitation, imagination, and the faculty of forming mental habits should be the objects of early education. Development of language skills and rote drill in arithmetic should be the principal and prominent object of the primary or common school. The higher principles of science can only be thoroughly understood by a mind more fully matured, for the judgment and the reasoning powers are of slower growth. Of serious consequence is the endeavor to invert the order of nature, and attempt to impart those things which cannot be taught at an early age, and to neglect those which at this period of life the mind is well adapted to receive.

P iaget? No, Joseph Henry, 150 years ago. As happens to all educational fads, the kernel of truth in the theory of concept development dies and is resurrected in every generation. Everyone knows that you can't teach a 6-month-old child to walk. You can lecture interactively, provide peer instruction, give multiple-choice tests. Beat the kid. Still won't walk. But if you wait half a year or so, you can't keep him from walking. So we shouldn't be surprised when the child can't tell analog clock time at the age of five. (Not your child, of course, who is precocious—but most children.) If you teach clock time in third grade, about half will learn and the other half will be convinced that they are stupid. If you wait until fourth grade they'll all know how to tell time without being taught. In Piaget's terms, this concrete operational stage lasts until puberty. Then the child becomes formally operational and can be baptized or bar mitzvahed or taught algebra.

The thirteen-year-old won't necessarily *learn* algebra, of course. Contrary to the expectations of the schools, about two-thirds of a standard population will never really understand how to find z if $x = y/z$. That's why the U.S. Navy in World War II taught Ohm's law in three forms.

So what do we pretend to teach in the schools? In the elementary-school science series of one major publisher, we find atoms presented in third grade, gas laws in fourth, Newton's laws in fifth, and nuclear fusion in sixth. In the seventh-grade text of another publisher the children, having attained the age of reason, are told about the Coriolis force (incorrectly, as it turns out). In high school most students study about DNA and genes in biology before they learn about molecular orbitals the next year in chemistry, which they memorize before they learn about energy in physics the following year, if there is a following year. Until the 1960s, calculus was a second-year college course for most science and engineering majors. Now, in many American high schools, selected groups of students in their senior year can start calculus, even though many are still having trouble with algebra, trig, and geometry.

But there is no cause for alarm. Nature is not mocked. Humans are still humans, and learn things at their own pace. What they learn to do when presented with concepts for which they are not ready is to memorize key words and algorithms. That's just what Joseph Henry said was appropriate for younger children. Unfortunately, it's a powerless sort of knowledge, though some students become skilled at answering familiar multiple-choice questions. If the test questions involve novel situations, such as with the Force Concept Inventory, then we get the startling results that the students apparently don't understand the basic concepts. We weep and blame it on our methods of teaching.

Just possibly, the fault lies in our expectations. Maybe the students aren't old enough yet. Maybe it takes several times around the topic, presented in many different modalities, to begin to understand. The consequences for engineers and science majors would be to start out slowly with the study of lots of phenomena and simple quantitative relationships. Mathematical analysis could be put off until the second or third college year. For high-school students and for people who will take only one year of physics, the phenomenological study is the only thing they will remember anyway. As for third graders, let's get those silly atom diagrams out of their books and let them memorize languages or the multiplication table, or even the names of more dinosaurs. Joseph Henry would approve. He thought that it was wrong to invert the order of nature.

October 1997

Part V TUTORIALS

P hysics is filled with paradoxes and seeming contradictions. Usually the resolution of one of these problems leads to a deeper understanding. Studying physics is not a matter of learning something completely. Physicists never quite master the simplest laws, which are often the most profound. There is always a new viewpoint, a new exception. What could be simpler than Ohm's law? But it is valid only with certain materials and under certain conditions.

This group of editorials consists of tutorials. I presumed to present *the* truth on certain topics. Readers were invited to write me letters if they disagreed with my interpretation, but few did. Here then are definitive views on error analysis, Newton's laws of dynamics, harmonics of vibrating systems, models and reality, circular motion, and other subtle subjects that need extra worrying.

The power of instruction is seldom of much efficacy, except in those happy dispositions where it is almost superfluous.
—Gibbon

I really understand the material; I just can't do the problems.

n Victorian literature it was usually some poor female who came to see the error of her ways. How prescient of her! How I wish that all writers of manuscripts for *The Physics Teacher* would come to similar recognition of this centerpiece of measurement. For, Brothers and Sisters, we all err.

Now, regrettably, most computer systems are shameless in this regard, and present data and graphs that are error free. The data points on the graphs are points, indicating infinite precision. The transistors flip and there is the best-fit line, complete with slope and intercept. It isn't so in the articles in *The Physical Review.* There the graphs are messy with giant error bars indicating the uncertainty in each datum. One might conclude that in the ascent from high school to college to research work, precision is lost and sloppiness multiplies. Or, one might conclude that our simple computer programs are lying.

The error, or uncertainty, in a measurement is not the difference between the student value and the handbook truth. That is a *discrepancy,* and may well be within the bounds of the uncertainty, or error. The error is usually not given by any simple formula such as the standard deviation read out from a hand calculator. Instead, judgment is called for in assessing the range of values within which the recorded measurement lies. That range may be determined by the nature of the measuring instrument (meter sticks do not read to microns), by the skill of the measurer (five-year-olds do not read to microns), by the needs of the measurer (the length of a 2 × 4 need not be read to a micron), by the nature of the thing being measured (there is no such thing as the diameter of a piece of chalk precise to one micron). As for the needs of the measurer, the golden rule is that if a thing is worth doing it is worth doing well enough for the purpose at hand, and it is surely silly and probably wrong to do it any better than that. Precision is expensive. Don't spend time making a 1 % measurement when 3 % will do.

Ah, but there are rules about how many measurements to make in order to reduce the error—five in some lab instructions, 10 in others. If you measure a length 10 times, each to three significant figures, can you not take the average to four significant figures? Maybe *you* can, but introductory students can't. They just keep reading the same number over again, plus or minus one, 10 times. You can begin gaining precision by multiple measurements if your measurement is limited by probability. To determine that, plot the values obtained for each datum. If you get a bell-shaped curve, you can start using fancier statistical rules. But freshmen and high-school students don't work like that. Nor should they.

The standard deviation is an artifact of probability statistics. The formulas for line fitting assume that each datum represents a Gaussian distribution. Introductory students don't need these subtle points. The size of the error, the uncertainty spread, should be determined by taking account of everything the student knows about the apparatus and the procedure. Then the student should assert: "I guarantee that any other measurement of this quantity, made by anyone else, using any other kind of instrument will be within these error limits." The student should learn how to combine errors of this sort, using the simple rule that in addition the absolute errors should be added; in multiplication or division the percentage or relative errors should be added. Finally, is the result a 1%, a 5%, or a 10% measurement.

These are good rules too for most articles in *The Physics Teacher*. When sending in a manuscript, show those error bars and check those significant figures. To err is human; to describe the error properly is sublime.

October 1999

T he time has come to confess the errors of our ways; nay, rather, to proclaim them. Naturally, we are not talking about mistakes. We do make mistakes occasionally, but prefer to discuss them privately. As for errors, we mean, of course, the uncertainties of experimental measurements. There are absolute errors (a strangely assertive term), and relative errors. If I measure a length of 10 m with an uncertainty of 1 m, then my absolute error is ± 1 m, and my fractional error is 1/10. I have made a 10% measurement. If you now claim that some other length is 100 m with an absolute error of $\pm$m, then you have made a 1% measurement, even though your absolute error is the same as mine.

How did you know that your absolute error is 1 m? The crucial word is that you "claimed" it. On your professional honor, you have assured us all that the true length is between 99 m and 101 m. It's possible to refine that assurance. Under certain, rather rare, circumstances you might claim that your absolute error was a standard deviation of a Gaussian (normal) distribution of values. You would then be promising us that about two-thirds of all subsequent measurements of that length would fall between 99 m and 101 m. By using such a sophisticated criterion, you would furthermore be implying that your errors were caused by a random process and that these errors are larger than any other kind of uncertainty in the measurement. For instance, you would be promising us that you have checked to make sure that your meterstick itself is accurate to much better than 1%. Evidently citing absolute error involves your personal integrity and should not be undertaken lightly.

But we still haven't explained how to determine the error. For instance, if you measure the width of a table top, should you repeat the measurement five times, take the average and then calculate the range? Why would you want to do such a silly thing? Do you have nothing better to do? If you made a mistake the first time, you'll probably repeat it the second time. To guard against mistakes, perhaps you should get someone else to make the same measurement, preferably with another instrument. As for your absolute error, that's a matter for your judgment. First, there may be uncertainties in the measuring instrument. For instance you cannot read a meter stick closer than about a fifth of a millimeter. Are you sure that the accuracy of the stick is as good as the scale divisions imply? Maybe the stick is old and has shrunk by a millimeter or so. Second, there may be uncertainties in the meaning of the thing being measured. Most table tops have rounded edges. What is the meaning of the width of such a surface? Third, there may be uncertainties caused by the inexperience, or lack of time, of the measurer. Don't trust a freshman to read a meter stick to a fifth of a millimeter if the student hasn't compensated for parallax.

This third factor in the size of errors is one that should be considered first. Why do you need the information? Is a 10% measurement good enough or must you know the value to ±1%? Precision usually takes time, and time is money. It would be foolish, if not wrong, to spend time or money getting precision that is not needed. That maxim applies to students as well as to full professors. Perhaps it applies more to students. The full professor is on his own and has lots of time. It's the student who is busy who shouldn't be led astray. Teach the student never to do a measurement or make a calculation without asking first, "For what purpose is this information needed?" Without knowing the answer to that question, you don't know what kind of a measuring device to get or how much time to spend. In short, your absolute error should depend on your purpose.

After you have decided how much time to spend and what instrument to use—in other words after you have chosen a 10% or a 1% measurement—how do you know what absolute error you have obtained? There are no general rules. The awful truth is that where your honor is concerned, you must use judgment and all the experience you have. If the object to be measured is constant in time, then it usually makes no sense to make repeated measurements. If the phenomenon is repetitive, such as the swing of a pendulum, then there may be some reason to take several measurements to help you gauge your technique and the system's consistency. A range about the average may or may not be your absolute error. On your honor you are also certifying the accuracy of the watch, the consistency of your reflexes, and the reasonableness of your measuring method. If you do measure something where the data seem to scatter, is it a random effect of measurement? Or is it perhaps that the system being measured cannot be reset to the same initial conditions each time? Your statement of error means different things in these two cases. If you think that the data are subject to random error, and you have reason to care enough to do the analysis, then plot the data distribution. Is it Gaussian? If so, why bless you, go ahead and use the standard deviation. But then you probably aren't a freshman. Almost no freshman lab experiment can justify the use of standard deviation.

Once you judge the absolute error in a particular measurement, you must often propagate the errors as you multiply or add values (see *Phys. Teach.* 21, 155 [1983]). For multiplication or division, add the percentage errors. For addition or subtraction, add the absolute errors. These are the rules of maximum pessimism. It is assumed that there is no cancellation of a positive error by a negative error. Note one important implication of these rules. If you measure one value to 10% and are multiplying it by another value, it is useless to spend time getting the second value to much better than a few percent. In other words, you don't win by multiplying a two significant figure number by one with three significant figures. Your product has only two.

The main rule about choosing the value of an absolute error is that there are no easy rules. It's a matter of judgment, experience, common sense, and even integrity. Oh, let us hope that all our students will learn the errors of their ways.

November 1983

Insignificant Figures

T here is a grand old saying that if a thing is worth doing, it's worth doing well. That's utter nonsense, of course; if a thing is worth doing, it's worth doing well enough for the purpose at hand, and it is probably wrong and certainly silly to do it any better than that. We physics teachers should be particularly sensitive about this point. Physics is often thought to be the science of precision—with the common implication that this means fussiness. There is a feeling that everything must be measured to n significant figures and that advanced research consists of measuring to $(n + 1)$. Students in school assume that a meter stick must always be read to the nearest fraction of a millimeter even though in the real world at home they may be satisfied with a tape measure.

Far from being concerned with fussiness, physics is supremely the science of common sense. The first question that anyone should ask before making a measurement is, ''For what purpose is the information needed?'' The answer to that question may mean that the measurement can be taken casually in a moment's time, or the answer may demand the work of a lifetime.

What is the area of your front yard? Without further information, there is no good answer to the question and no reasonable procedure of measurement. Do you want to know the area of your front yard in order to buy some lime to put on the grass? Then you can measure the area of your front yard by glancing at it as you drive down the driveway on your way to the hardware store. Lime comes cheaply in eighty-pound bags and cannot be purchased in half bags or tenth bags. If you have a little too much or too little, the grass will never know the difference. Do you want to know the area of your front yard in order to buy some expensive grass seed for it? Then the measurement must be done differently although you would certanly not use a meter stick. In this case, the proper unit of measure is your pace—the double step, which for all adult males is a little over 5 ft (hence, our world mile which comes from the Latin, mille, a thousand paces); so, in order to get the right amount of expensive grass seed, you might pace out your front lawn keeping track of the numbers on the back of an old envelope if the lot is irregular in shape.

Do you want to know the area of your front lawn for tax purposes? For such a purpose, entirely different procedures and instruments are required. Get out the surveyor's chain and transit.

Particularly in the introductory course of physics, most problems should be concerned with functional dependence of variables and the order of magnitude of quantities. The posing of Fermi questions should be a class routine. Fermi delighted in finding approximate quantitative answers to seemingly impossible questions. How many piano tuners in New York City? Well, to one significant figure, how many people in New York City? how many families? how many pianos per family? how many pianos can a piano tuner tune in one day? how many times a year is a piano tuned? how many tuners are needed? Try the numbers yourself and, by placing limits on your uncertainties, decide whether or not your answer must surely be accurate to within a factor of ten in either direction. How fast can a football player run? No one is concerned with precision here. Guess the time that it takes to run fifty yards and bracket your guess with the percentage of uncertainty. Perhaps you can get an answer to within a factor of two, or if you are familiar with football, to within 40% or 20%. To one significant figure, what is the average thermal energy of a gas molecule at room temperature? How does this compare with the gravitational potential energy of that molecule on the surface of the earth or with the chemical binding energy of most common molecules? One significant figure is all you need to find out whether or not the molecule will escape from the earth or will be destructive to ordinary compounds. Which has more influence on the tide—the sun or the moon? One significant figure in the ratio is all we need to make the point.

Of course there are situations where precision is vital, where a model fails to agree with the experimental data only after many significant figures have been calculated and measured. Very small discrepancies between theoretical model and experimental data have often revealed new and important phenomena. Pluto was discovered because of very small discrepancies in the orbits of Neptune and Uranus. The very small Lamb shift in the atomic spectrum of hydrogen led to a new view of quantum electrodynamics. Dicke at Princeton and others have gone to a great deal of trouble to find that, to eleven significant figures, gravitational mass is equal to inertial mass. Fussy? No. The general theory of relativity is at stake.

Unfortunately for students, but fortunately for the lure of the game, there are no rules to tell you what precision is needed in any particular case. There are statistical rules, often misused, for the analysis of errors, but only judgment, experience, and uncommon common sense can tell you when precision is required and when it is not. The sin of requiring too much precision is as deadly as that of being satisfied with sloppiness. To avoid excessive sinning in either direction, we should enshrine in every classroom and laboratory the watchword, ''For what purpose is this information required?''

March 1968

On Certain Errors and Uncertain Virtues

I n our continuing study of the theological aspects of physics teaching, we consider today the comparative venialness of two common sins. Which is more to be condemned in experimental work and reporting—sloppiness or undue fussiness?

We see the results of both in journal reports and manuscripts. In a recent issue of a sister magazine, an author claimed that his experiment with a particular apparatus confirmed a standard law. His published data curve went straight as an arrow through the origin. But there were no error bars on the data points. Indeed, there was no discussion at all of possible error. It was a special pity because the phenomenon described could not possibly have produced a straight line with this apparatus.

On the other hand, about once a month we get a manuscript where no detail has been considered too small to mention. Every measurement is made and cited to four significant figures, and the reader is expected to follow every turn of the micrometer. This style of lab report is also common in the *Physical Review* and the *Journal of Research in Science Teaching.*

Like many vices, these aberrations are really virtues only slightly twisted. There is a time for free wheeling, crude, approximate, order-of-magnitude work, and there are other times for meticulous, detailed, precision work. Experimental skill consists of using the right method at the appropriate time. Part of reportorial skill consists of justifying in advance the need for precision before describing how it was obtained.

As we all know, error analysis, or rather, uncertainty analysis, seems to be one of the most difficult tasks for beginning students. They want rules. "Instruments should be read to the nearest one-half (or is it one-fifth?) of the smallest scale division." "Take the reading ten times, and then the average can be cited to an extra significant figure." "All data should be given to four significant figures—or as many as the calculator can generate." The frightening truth, of course, is that error analysis is seldom subject to general rules. Instead, uncommon common sense must be used.

Suppose you time a single period of a pendulum. If the period is about one second, and you are timing with a clock that you control manually, your uncertainty (the error) is almost certainly about 20%. If you average ten readings, your percentage error does not change. There is nothing random about the type of human error involved, and thus there is no assurance that there is any statistical cancellation. On the other hand, if you measure the time for ten swings, then your absolute error is still about one-fifth second. Your percentage error is thus reduced to about 2%. Even these generalizations must be qualified. Are you sure that the period of a pendulum is sufficiently independent of amplitude that you can add together ten swings, since the amplitude of the first is greater than that of the tenth? (The period of a pendulum does depend on the amplitude, and if the swing is greater than 30°, the effect will be noticeable.) How do you determine that your clock and your technique are really good to one-fifth second?

If you measure the position as a function of time of an air track glider, how do you deal with the possible errors in measuring Δx versus time? A plot of $\Delta x/\Delta t$, versus t, is a graph of $v(t)$. Use error bars to indicate the uncertainty in each value of $\Delta x/\Delta t$. Then find the acceleration and the error in the acceleration by testing whether or not a straight line can be drawn through the error bar regions. You can find the possible range of values of the acceleration by measuring the maximum and minimum slope of a line that can pass through all the data regions. If the acceleration was not constant, that fact will be displayed in the graph. In fact, the best way to analyze error in almost any kind of functional relationship is to plot the data with error bars. Each error bar must represent your considered guarantee that the value lies somewhere within the error bar region. Any data presented without such analysis is automatically suspect.

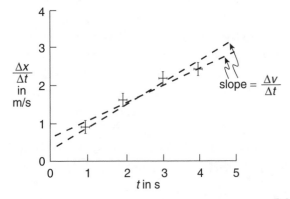

Of course, before you start taking data and trying to judge the uncertainty, you should determine in advance why you want the data and how precise they must be. Precision is usually expensive. To pay for it, when you don't need it, is wrong. To report data to high precision without explaining why you went to the trouble is as suspect as reporting the data without any information about errors.

To avoid error, of the type that is wicked in our profession, first explain why you need the precision you are aiming for. Then present the data and the uncertainties graphically and send it off to *The Physics Teacher*. So may your days be blessed, your students grow in wisdom, and your paper be published.

November 1975

On the Sizes of Things

S o this kid comes to me and wants help with a homework problem in relativity. The data are in kilometers per hour and I suggest that the first thing to do is to change it to meters per second. He comes back in a few minutes with the speed of the rocket ship being 4.86×10^9 m/s. "Something's wrong," I say. "That can't be right." "How do you know that?" he says. "What's the speed of light," I ask. "I'll look it up," he answers.

Now we pride ourselves in Physics in not having to memorize a lot of things. We are not like biologists or organic chemists who must be able to spell sentence-long names of complicated molecules. On the other hand, there are some facts that should be learned at your mother's knee and never forgotten. Of course, there are also facts that need be remembered only until after the final exam in the course. Let's consider only the important ones. I'll suggest my list; you can make your own.

Naturally, everyone should know that $c = 3 \times 10^8$ m/s. It's not only the speed of light; it's the ultimate speed. While of less importance, but often useful, is the value for the speed of sound in air: $v = 340$ m/s. Once you know that number you can calculate the wavelength of the 440-Hz note that the orchestra tunes to and can make educated explanations about diffraction of music around furniture and people.

The electron volt is the unit of subatomic energy. Witness the fact that all electrochemical cells yield a volt or two (1 1/2 for the D cell, 2 for each cell in the car battery). Therefore, most chemical reactions involve an electron-volt of energy, including the photographic process, sun tanning, and photosynthesis. So the energy of a visible light photon must be a couple of electron volts.

To link the micro- and the macroworld, we should know the size of the mole. For physicists it's the name of a big number equal to 6×10^{23}. Once a student told me that he knew it was 10^{23} but he didn't know whether it was $+$ or -23. It's $+$. One really ought to know, at least for the duration of the course, that the unit of electric charge is 1.6×10^{-19} coulomb. That's $-$. Of course, if you're dealing with the microworld, you have to know that Planck's constant is 6.6×10^{-34} J·s. You can then figure out, or know intuitively, that the wavelength of visible light is about half a micron.

Anyone taking physics learns that $g = 9.8$ m/s² $= 9.8$ N/kg. For most calculations $g = 10$ is good enough. It can be a touchstone for gauging the sizes of other accelerations. A human accelerated to $10\,g$ is dead. It's useful to know that the speed of a brisk walk is 4 mph, which, divided by 2, is 2 m/s. While we're on homely examples, consider that a pint's a pound, the world around. Similarly, a liter's a kilogram in every land.

The circumference of the Earth must be about 24,000 miles because there are 24 hour-long time zones around the Earth, each one of them about 1,000 miles wide. (Consider that it's about 3,000 miles across the U.S. using up three time zones. It must be about 3,000 miles because a jet liner takes 6 hours at 500 mph.) It must be about $[(3 \times 10^8$ m/s) (8 min) (60 s/min)$] = 1.5 \times 10^8$ km $= 90 \times 10^6$ miles from Earth to Sun, since the sun is 8 light-minutes from us.

The resting heart beat of an adult takes about 1 second—half that for a newborn. That's the time it takes to recite "one thousand and one, one thousand and two. . . ." Each syllable takes 0.2 second. You can measure g this way by timing a falling stone, good to 20%. Everyone should know their pace, the double step. For an adult, the pace is about 5 feet, hence comes our word "mile" from the Latin word "mille" for 1000.

As for magnitude of power, our houses are filled with touchstones. The toaster consumes 1,000 watts, and since it is being provided by 120 volts, the current must be 8 amperes (amps). The cable connections to the car battery are very thick, presumably to carry large current with little energy loss. At 12 volts, 60 amps will provide 720 watts, enough to power the starting motor rated at 1 horsepower.

My list of such familiar magnitudes could go on and on. Yours probably could too, although there would be some differences. Along with Newton's laws in the first semester, and Ohm's in the second, students in the introductory course in physics should get experience in the relative sizes of things. Numbers really do count.

To a First Approximation

Fourth week of classes and this student comes up to me and says, "Why don't you just tell us the truth?" Well, he looks sincere, and I say, "Sure, what subject did you have in mind?" "About physics," he says. "You keep telling us that what we're learning is only a first approximation. When do we learn the truth?"

Not in this world, friend. Or at least, not in this department. The best we can do in physics is to work our way through a series of successive approximations. Consider how we explain such a simple phenomenon as the flight of a projectile. The vertical component of velocity acts independently of the horizontal component (except, of course, if it has wings). We assume that the horizontal component remains unchanged during the flight, following Newton's first law (which applies to and defines an inertial reference system, which the earth is not). The vertical component of velocity is subject to constant acceleration (which would be true if the earth were flat, nonrotating, and without air). Still, the familiar formulas provide a good first approximation for baseball throws, if not for rifle bullets or ping-pong balls. First-year students have enough trouble with the first approximation. The second approximations would challenge upperclassmen. Absolute truth is a problem for National Aeronautics and Space Administration (NASA).

At least with Newton's second law we're on solid ground. $F = ma$, all during the first semester. But not quite. Actually, $F = d$ (momentum)$/dt$, and momentum is only approximately equal to mv. Even when it is equal, you can't always remove the m from the derivative, for instance, in rocket problems.

You can depend on friction, though. As is well known, the coefficient of dry sliding friction is independent of velocity. There are, however, two values in the handbook for each pair of materials—one for zero velocity, and the other for all other velocities. Is there a transition zone? Except for that, is the coefficient independent of velocity? Well, almost. Within limits, of course.

We don't even pretend that real gases are ideal, though the rules that we develop are for ideal gases. If you're going to be fussy about $PV = nRT$, try $(P + a/V^2)(V - b) = nRT$, or one of the other approximations.

At least for our first-year problems with mixture of materials at different temperatures, the specific heats stay constant. Not that they do in the real world, of course. In fact, if we examine the next approximations about specific heats we are suddenly in the realm of quantum mechanics.

You know how to derive Bernoulli's equation. That's the one about the fluid in a pipe with different cross sections and heights. The pressure along the pipe is constant as long as the velocity and height are constant. Unfortunately, it's hard to maintain a pressure head with a hose like that, even to first approximation. Of course, we don't really mean a hose. We're talking about a boundary of stream lines, as all of your students clearly understand.

It's certainly convenient that spherical lenses yield such a simple focusing law. Naturally, the rays must be paraxial. If they aren't, the equation is not child's play. In fact, there isn't really an equation for the next approximation.

Do your students still think that the period of a simple pendulum is independent of its amplitude? Why not jump immediately to the full solution? For that matter, why Bohr them with a simple atomic model when you can start right out with Schroedinger's equation? That's what the chemists do.

The truth is that there is no simple truth. We comprehend the world with approximate models and refine them with successively more complicated ones. The game is obvious in physics, less so in other fields. The students themselves pass through these stages. In their case, as with physics, we should not be surprised by the crudeness of the first approximation but only hope that the series is converging.

November 1980

Models and Reality

This is the season of lights in darkness and new beginnings for old dreams. It is the season of metaphors. Physicists understand metaphors and models. We use them all the time. Consider our December centerfold where we portray models of a falling human being.

If you want to analyze the fall of a human (physically through space, not spiritually through error), it's awkward to get volunteers for a long drop. Our centerfold shows pictures of people falling, but note that for us, the viewers, these are only two-dimensional models of the real thing. To carry out an experiment with humans would require not only volunteers but fairly elaborate apparatus. Physicists know what to do when it is expensive or complicated to deal with the real thing. We simplify the problem and make an approximation. We choose a model whose behavior is simple enough so that we can measure and calculate it. Then we analyze the relationship between our model and reality.

In an introductory physics class, our model of a falling human might well be a golf ball. It has no arms to flail, the center of mass is easy to locate, and no harm is done when it bounces. Of course, the shape, the ratio of weight to surface area, and the size of the ball are different from those of the human, but the golf ball will do for a first approximation. If you just look at a falling golf ball, it is not easy to analyze what's happening. It appears to go faster and faster, but only if you're looking for that effect and expect to observe it. As physics teachers, we know how to make the motion more apparent. Take some of the information away and observe the ball's position only at selected times spaced at equal intervals. We show such a record (a model of a time exposure) in the centerfold. Even that information does not reveal as much as does the more abstract model of a graph of $y(t)$. That parabolic graph does not look much like a falling human, but nevertheless it allows us to describe the location (perhaps of the body) at various times. The graph itself can be described by another model, the algebraic formula underneath the graph. People familiar with the use of such mathematical models can then easily describe an important aspect of the fall of a human. From the original photograph, or from the graph of $y(t)$, we can next construct another graphical model showing $v(t)$. From this model it is apparent that the falling object is going faster and faster in a very specific way; its speed is proportional to the time of fall. All these pictures, graphs, and formulas are models of a falling human being.

Models always have a limited range of application. Our simple model for the fall of a human does not account for air resistance. We can expand the model, of course, either through experiment or by using theory derived from other models. Air friction for a golf ball, or a human, is proportional to the square of the speed. Our centerfold shows a graph of $v(t)$ for such an expanded model. Is our more primitive model wrong? No, for short fall times, where the speed is still well below terminal velocity, the simple model is easier to use. In most cases it tells us all we want to know within the precision we need. Note that after long fall times, we again can use a simple model; the speed is constant. The complicated model must be used only during the transition period.

We present one other model of a falling human. It is a fragment of a poem by James Dickey that was published in *The New Yorker* many years ago. Dickey took as his theme a brief news item from *The New York Times* that described the death of a stewardess who had fallen from an airplane. The poem is very long and accurately describes many physical aspects of the fall of a human being, as well as other aspects that are not touched upon by our other models.

Does physics deal with reality or only with models? Is it reality that there are electrons *in* atoms, which we can knock out, leaving ions? Is it reality that there are *not* photons as such *in* atoms, but that we create them in the process of disturbing the atom and kicking them out? Certainly these are the processes according to our atomic model, but in the experimental reality we raise the temperature of a filament and out come both photons and electrons. Is it then reality that in a similar, more energetic excitation, electrons are created when they are ejected from the nucleus, but they were not in there to begin with? Certainly that is the truth according to our very successful model of the nucleus.

With only a few exceptions, when we see stars we see light only from point sources. We "see" no disk, no diameter, no list of elements, no rotation, no speedometer reading. But in many cases we can tell all these things by applying our successful models concerning spectral intensities, frequency shifts, and stellar formation. Many of the models are based on our successful models of the atom. Indeed, our present models of the origins of the universe are interwoven with our models of subatomic particles.

Are physicists peculiar, dealing with models as if they were reality? What do you suppose economists do, interpreting trends of the economy? What do you suppose each of us does, dealing with students, or they with us? Do we know the complete reality of another human or do we interact on the basis of a sexist, or racist, or Piagetian, or even a Freudian model? We all interpret reality in terms of models. Frequently the model is our reality.

Did a young woman really fall to her death over Kansas long years ago? That abstract fact about a falling human being has no emotional reality for me. It is just something that I read in *The New Yorker,* which quoted *The New York Times.* But the reality is that I cannot read Dickey's poem without a lump in my throat. For me, in this case, the model has become the reality.

In physics laboratories we deal on one level with very real meter sticks, wires, and lenses. In many cases we can describe the phenomena with formulas that can be learned by rote. It's the same as it is in our everyday life where the sun rises (in spite of the Copernican model), we know how to make the toaster work (with its electrons moving back and forth), and we can leave for school with reasonable good will to the rest of the family (having dealt with our interpretation of how they were feeling). Most days and in most circumstances we don't really have to worry about the nature of things or the reality of models. It can be pretty boring. How fortunate we all are to have a season that lifts us out of our everyday lives, a season when metaphors of the human condition are embodied in song and family ritual. How fortunate we physics teachers are to be practicing and teaching a discipline where models and reality are so intertwined that we cannot really understand one without knowing the nature of the other.

December 1989

W e've been reviewing introductory physics texts lately. Naturally, if we were writing the books we would present topics differently, and so would you. None of us is ever really satisfied with a text we haven't written ourselves.

One of my quibbles has to do with the teaching of Newton's three laws. Now, of course, Newton has many laws. There's dynamics, and optics, and gravitation, and cooling. But when we say "Newton's three," everyone knows we mean the trinity of dynamics. What concerns me is that many texts seem to trivialize these, making the first a special case of the second, defining mass with a wave of the hand as the quantity of matter, and for the third law creating two forces known as action and reaction.

Here's a standard text definition of Newton's first law. "An object at rest remains at rest, and an object in motion continues in motion with constant velocity unless it experiences a net external force." That seems reasonable enough. However, it seems like a special case of the second law; if $F_{net} = 0$, then $a = 0$. Why do we need the first?

An astronaut leaves his coffee tube in midair. With respect to the astronaut, the tube's acceleration is zero; hence, by Newton's first law the net force acting on the tube must be zero. I hold a plumb line as my jet plane starts its run. The bob points at an angle down and back and is motionless. The bob must be subject to zero net force, made up of tension in the line in one direction and some invisible action-at-a-distance force in the opposite direction.

"Ah, no," you say. "In the satellite the coffee tube is subject to a gravitational force that provides the centripetal acceleration required to maintain the tube in its orbit." "All I know is," replies the astronaut, "the tube is not accelerating with respect to me, and I'm a trained observer. If I agree that the Earth is exerting a force downward on the tube, then evidently there must be an equal and opposite force pulling the tube outward. That centrifugal force balances the gravitational force, leaving the tube motionless as required by Newton's first law." "Nonsense," you say. "A centrifugal force is fictitious."

We could have the same conversation concerning my plumb bob in the jet plane. Plumb bobs point down, which manifestly is partly in the direction back toward me. You explain, "The tension in the line supports the weight of the bob down and also provides the force necessary to accelerate the bob forward." "What acceleration?" I ask. "The bob is motionless with respect to me, and I am a trained observer. This is clearly a case of Newton's first law."

These are reference-frame problems. In the satellite there is a genuine force in the outward radial direction. In the jet plane there is a very real force shoving me back into the seat. The function of Newton's first law is to establish a reference frame for the second and third laws. If the reference frame is rotating, then we have "inertial" force fields with special names of centrifugal and Coriolis. If the reference frame is at rest, then we have fewer force fields, although calculations describing the resulting phenomena may be more difficult. Did I say *at rest?* At rest with respect to what? After all, who are we here on Earth to chide the astronaut about being in a noninertial reference frame? The astronaut is only revolving; we're spinning!

How can we determine that the net force on an object is zero, except by accepting Newton's first law and looking for acceleration? We could isolate the object from obvious contact forces by suspending it (somehow) in a vacuum. Then we could isolate the object from electromagnetic forces by surrounding it with an electrically conducting shield (a Faraday cage). The only long-range forces we know about are electromagnetism and gravitation. As a practical matter, how will we isolate the object from gravitation? How will we know that the object—like the astronaut's coffee tube—is not subject to "inertial" fields, which Einstein assures us are completely equivalent (at any point) to gravitational fields, which, after all, are curvatures in space. (This assertion is the cornerstone of the general theory of relativity, which is our theory for that fictitious, or at least mysterious, force known as gravitation.) Can we escape this dilemma by defining an inertial reference frame as one where Newton's laws hold if the only forces providing action-at-a-distance are electromagnetism and gravitation? If so, then Newton's laws are valid in an inertial reference frame, which is one in which Newton's laws are valid.

If you don't like that go-around, then how about this for a solution?

Let's define Newton's first law for a reference system that is at rest (or moving at constant velocity) with respect to the fixed stars. What fixed stars? Actually, it's practical to use the old-fashioned idea of fixed stars, even though we know that we and all the fixed stars are moving around in a galaxy that is moving around neighboring galaxies with everybody fleeing from everyone else. Galactic motion is just a philosophical argument as far as Newton's first law is concerned. Perhaps these days we have an even better solution as a result of the measurements made by the Cosmic Background Explorer (COBE) satellite. The 2.7-K electromagnetic background radiation is indeed fixed, or at least almost uniform in all directions, and we mortals (or anyone else) can measure our motion relative to it.

So we have an operational way of defining zero force and in the process have established our reference frame. Next we must find the effects of a nonzero force, preferably without getting into circular reasoning. That's the second part of the trinity, and sure we'll "worry" it in another editorial.

October 1998

What could be simpler? Exert a force on a mass, and it accelerates, $F = ma$! You can work that formula into any number of problems about objects on inclined planes, or Atwood's machine, or blocks sliding along on a surface with coefficient of friction, μ. What took Newton so long to discover his second law?

For one thing, the formula doesn't describe everyday experience. When you floor the accelerator of your car, you soon stop accelerating and just travel along at some constant, maximum speed. When you push a heavy box it takes a threshold force to get it moving, and even then it usually doesn't accelerate. If you want a Newton's law that matches the real world, try his fourth: *the weight of an object is proportional to the time you carry it.* Everyone who has ever carried a suitcase from one end of an airline terminal to the other knows the validity of that law.

Actually, there are so many problems with Newton's second law that one hardly knows where to begin. Newton had the same difficulty. There is the problem of how to define zero force. In the October editorial, I pointed out that to define zero force you had to define the reference frame for your measurements. Newton chose a reference frame at rest or in constant velocity with respect to the fixed stars. We can do no better, except, perhaps, by choosing the 2.7-K background radiation as our fixed reference. As a practical matter, the fixed stars will do just fine.

In our fixed (inertial) reference frame, we know how to make operational definitions of displacement, velocity, and acceleration. Now, to complete the second law, do we define F and so pin down m, or do we start with m? Let's start there. Everyone knows what m is; it's the quantity of matter. How do you measure that quantity? According to one misguided way of describing such a method, you *mass* it with a pan balance. Aside from the lamentable transformation of a noun into a verb, this procedure involves a *weight* comparison. Weight arises from the gravitational attraction of two objects, which has no prima facie relationship to the inertial quantity of matter that is involved in Newton's second law. Unfortunately, we use the same symbol for mass in both the laws of dynamics and the law of gravitation—both identified as Newton's laws.

$$F = ma \quad \text{and} \quad F = Gm \frac{M}{R^2} = mg$$

No wonder our students are brainwashed into thinking that inertial mass, m_I, and gravitational mass, m_G, are the same thing. Actually, as one of the cornerstones of the general theory of relativity, it turns out that they are the same thing: $m_I \equiv m_G$.

Galileo's suggestion about dropping weights from a leaning tower could have tested this equality to about one part in 10^2. Newton assured himself of the equality by timing pendulums made of different materials, and concluded that he could have noticed a difference of one part in 10^3. Eötvös, and more recently Dicke at Princeton, tested the identity to within one part in 10^9.

So we could, and we do, set up standards and compare masses on the basis that inertial and gravitational mass are identical. It would be a shame, however, to let the identity of symbols for mass hide the profound revelation that the gravitational attraction between two objects is linked with the reluctance of those objects to change their momentum.

But there are more mysteries involved in our simple quest for the meaning of mass. Mass, after all, is the same as energy. It is not a matter of one turning into the other; they are the same thing, though we often use different units for each. The mass of the proton is 1.6724×10^{-27} kg. The mass of the neutron is 1.6748×10^{-27} kg. The sum of those numbers is 3.3472×10^{-27} kg, but the mass of the deuteron is only 3.3431×10^{-27} kg. That's a difference of 2.3 MeV. The missing mass did not turn into energy. A gamma ray was emitted, and a gamma ray is not energy, although it transmits energy.

Let's examine Newton's second law again by starting with a set of definitions of force. The primitive experience of forces is that they can do two things—accelerate objects and distort objects. Although the second law is concerned with dynamics, we could start first with statics. Define a standard force in terms of the pull of a standard spring stretched a set amount. In no way can we call upon Hooke's law to provide a scale of forces based on this standard. Hooke's law, which applies under certain conditions to some springs, must be examined later experimentally. But we could make a set of physically identical standard springs and test them against each other. We could define double force as the pull of two such springs in parallel, finding consistency with several such combinations. Then we could explore the vector nature of force, which is not something for definition, but for experiment. Having thus set up a calibrated force system, we could turn once again to dynamics. Does constant force on an object produce constant acceleration? That's something to be determined by experiment.

We have barely touched on the richness of Newton's second law. Wouldn't it be dreadful if there were nothing more to $F = ma$ than a rubric for finding the answer on a homework problem? It seems to me that if physics students haven't been given a glimpse of the significance of that law, or haven't had to puzzle over some of the complications that are buried in it, they have been cheated.

November 1998

n the October editorial (Trinity Sure—I) we asserted that the role of Newton's first law is to establish a reference frame in which the second and third law make consistent sense. We defined such an inertial reference frame to be one that is at rest or moves with constant velocity with respect to the fixed stars. The November editorial worried Newton's second law and proposed methods of defining mass and of setting up a force scale. Experiment (not definition) shows that forces can be represented by vectors. Experiments also show that inertial mass and gravitational mass are equal, and in Einstein's view, identical.

However, there are still more mysteries wrapped up in Newton's third law. So far, we have considered the effect of a force acting on an object. Where does the force come from? What if one part of a compound object exerts a force on the other part?

Remember that we defined inertial mass in terms of gravitational mass, using a pan balance. (That's the method used with the international standard kilogram.) Another way to set up a mass scale is to run a series of collisions between objects on a frictionless track. Let an object with unit mass (any convenient standard) collide elastically (perhaps repelled by magnetic fields) with a physically identical object. The collision could also be an explosion driving the two apart. We observe that the velocity of the center of mass of the identical objects is conserved. A mass scale can then be set up by declaring that in such collisions the velocity of the center of mass always remains constant, thus defining the masses of the target objects as multiples of the unit mass. For instance, in an explosion between standard and test object, if the standard object moves away with $v_s = 2$ m/s and the test object recoils with $v_T = 1$ m/s, then we define the mass of the test object to be twice the unit mass. We are asserting that a new quantity is conserved when isolated objects interact—momentum. For speeds much less than that of light, momentum is defined as the product of inertial mass and velocity. In explosions $(m_T \vec{v}_T) = -(m_s \vec{v}_s)$. In internal collisions, the momentum of an isolated system remains constant.

This quantity called momentum, which starts out being defined as a product of two familiar quantities, ends up being more fundamental than either quantity separately. There are situations, particularly with subatomic particles, where interactions can be described in terms of momentum exchange, and the mass and velocity are either not measurable or not meaningful. Furthermore, it turns out that

$$F = ma = m \frac{dv}{dt}$$

is just an approximation. The more general second law is

$$F = \frac{dp}{dt}$$

where p is momentum.

Consider one consequence of the expanded definition of the second law. For rockets, the velocity of the exhaust gases with respect to the rocket remains constant.

Therefore,

$$F = \frac{d(mv)}{dt} = v \frac{dm}{dt} \ .$$

The thrust is equal to the product of exhaust velocity and the time rate of ejection of mass.

Newton's third law is best characterized as being equivalent to the conservation of momentum. Consider two objects repelling each other.

To conserve momentum during any time interval, Δt, $\Delta (\vec{mv})_1 = -\Delta (\vec{mv})_2$. Therefore, $F_{2 \text{ on } 1} \Delta t = -F_{1 \text{ on } 2} \Delta t$, and $F_{2 \text{ on } 1} = -F_{1 \text{ on } 2}$. Considering this an isolated system, the internal forces cancel out.

One hallowed but unfortunate way of describing this situation is to say that action equals reaction. Have you ever seen an action meter? In the stock market, action can lead to reaction, but in physics, action is defined as $\int p\,dx$ and has nothing to do with Newton's third law.

Another popular way of describing the third law is to claim that forces occur in pairs. That can be misleading. The forces are not two separate things; they are a single interaction. To be sure, with the second law we isolate a particular object and account only for the external forces acting on that object. With this third law, we have a system of at least two parts and describe the internal interactions that produce forces on the individual parts.

Of course there are still more complications with the definition of mass and force. For instance, in a system that can be described in terms of potential energy, the system can exert a restoring force:

$$F_x = -\frac{\partial U}{\partial x}.$$

For another example, forces not only add like vectors, but also follow vector rules for multiplication.

It's generally true that the simplest things are most profound. The three laws of dynamics didn't spring full-blown from Newton's head. He wrestled with them, using terms and concepts less facile than our modern ones. In our teaching we must, of course, demonstrate how to solve dynamics problems using Newton's three laws. But we should also exploit the opportunity to give our students a hint of the complexities and fascination of these simple and fundamental laws.

December 1998

Revealed Truth by the Editor on Certain Subtle Points

Strange things cross the editor's desk. About twice a week we discover something that we had not known before. Also about twice a week, filled with our newly found wisdom, we discover that somebody else doesn't yet know these things. Sometimes the matters are really technical in nature. Usually these concern principles that we learned in high school and have never really thought through since. Often, however, the problem is more one of semantics. The language of yesterday confuses the interpretations of today.

We have accumulated a list of topics that still cause subtle confusion. From time to time we will use this column to set things straight, *ex universitate* if not *ex cathedra*. If you do not like any of our arbitrary pronouncements, write us arbitrary letters.

First of all, there are these books and articles entitled *Matter and Energy.* What's the matter with that? Aren't those the two great divisions of things? *No.* There is no good definition of matter, except perhaps that it is something to stub your toe on. For scientific purposes the term is an anachronism. Is matter something that occupies space and has mass? So does an electromagnetic wave. A photon is every bit as good a particle as is a proton. It is true that photons and neutrinos do not have *rest* mass, but they do have spin and carry momentum and energy (hence, mass).

All too often matter and mass come to be used interchangeably, so that Einstein's famous equation is described in terms of *matter* turning into energy. The m in mc^2 stands for *mass,* of course, and mass does not turn into energy. Mass and energy are the same thing, though usually given in different units. Einstein's equation merely gives the conversion factor between kilograms and joules, even as we might give a conversion factor between inches and feet.

Is it not the case that when a positron and electron annihilate, they turn into energy in the form of two photons? *No.* Photons are not energy any more than man is a voice. Photons carry energy just like protons do and just like men (or women) have voices. It's one of their attributes.

Even if matter could be rigorously and usefully defined, why divide the world up into matter and *energy.* Why not matter and *love?* Or

Cliff's Nodes

matter and *organization?* Perhaps these two are out of favor because love is notoriously a nonconserved property and organization gets us into the second law of thermodynamics. If we insist on using energy as one of the great dichotomies, why not be alliterative and speak of *matter* and *mass?* Surely no one would deny the equivalence.

The truth of the matter is that some people still think that energy is a substance, something like caloric. The implication is that it is in some way opposed to matter, or at least opposite. It is a strange hangup to have, since we have known about energy as a conserved property only for the last 130 years. It seems unfair to single out this particular conservation law when there are a number of others just as important and with a longer claim on our study. How about *matter* and *momentum,* or *matter* and *electric charge,* or *matter* and *baryon number?* Those are all good conserved properties. For the introductory course where we need not mention the weak interaction, we could consider *matter* and *strangeness.*

If the world must be subdivided, it would make more sense to consider the particles (protons, mesons, electrons, photons, etc.) communicating with each other through four types of interaction (strong nuclear, weak, gravitational, and electromagnetic), subject to a variety of laws (conservation of energy, momentum, etc.). These divisions should not be considered in too rigid a fashion since some of the particles are to some extent agents of the interactions and all of the particles may be to some extent creatures of each other.

It would be better not to try to subdivide our subject like this at all. Students wouldn't have to memorize silly distinctions, which waste a lot of energy and don't really matter.

October 1972

More Revealed Truth: On Going in Circles

I n line with our fearless policy of saying sooth on certain subtle aspects of physics teaching, we shall now tackle the thorniest, if not the most vital, question in modern physics: Is the radial force on an object in circular motion centrifugal or centripetal?

There is no question about the magnitude of the centripetal–centrifugal force. It is equal to mv^2/r or $m\omega^2 r$. The question is, what is its direction? Is it a center-fleeing force (centri*fugal*), or one that impels toward the center (centri*petal*)? There are those who point out (correctly) that, according to Newton's second law, an object in motion will continue moving in the same direction with constant speed unless acted upon by an external force. Therefore, if an object is to travel in a circular path, it must be operated on by a centripetal force continually changing the object's direction *inward* radially, hence making it deviate from a tangential path. There are others who assert (correctly) that if you are in a rotating system you certainly experience a force tending to throw you *out* radially. For instance, if you are in the passenger seat of a car turning sharply toward the left, you are in danger of being thrown out of the car to the right if the door on your side is not securely fastened. The coroner would hesitate to say that you had been the victim of a fictitious force! The roots of plants grown on turntables follow the direction of the vector sum of gravity and centri*fugal* force. Plumb bobs on merry-go-rounds also indicate that their local "down" direction has an outward radial component. (See Bartlett's article on "Which Way Is Up?" *Phys. Teach.* 10, 429 [1972].)

The fictitious-force man would argue that these centrifugal effects are easily explained by looking down on the rotating system from a position that is not rotating. The person thrown out of the car has just been released from the centri*petal* force that had been causing him to go in a circular path, and now hurtles out of the car tangentially to the original path. (The new path looks radial to a person in the car.) The roots and the plumb bobs on the rotating system are fastened to flexible constraints that allow them to swing out to larger radii until the radial components of the constraints provide sufficient centri*petal* force. Can these two viewpoints ever be reconciled?

Consider two similar but simpler cases. If a person is riding in a car at constant speed and drops a ball to the floor, what is its true path? The person in the car could declare that, whether measured by eye or measured with instruments, the path of the falling ball was a vertical straight line. He is right. An equally proficient observer stationed beside the road would announce that the path of the ball was parabolic, since it was accelerated downward and had an initial horizontal speed that remained constant. He is also right. The "path" depends on the reference frame from which it is observed. Suppose that a person is in a jet airplane during the initial acceleration down the runway. He is subjected to some very tangible forces shoving him *backward* into the seat. The seat cushions compress; the seatbelt feels looser. A plumb bob (or a magazine suspended from a corner between two fingers) will confirm the fact that objects in the plane are experiencing a force toward the rear. But what would an observer outside the plane measure? Clearly, everything in the plane is accelerating in the forward direction and, hence, must be subject to a force in that direction. That observer is also right.

There is nothing paradoxical or mysterious about this at all. Both the magnitude and direction of forces depend on the reference frames in which they are measured. Nor is there anything sacred about one particular reference frame. "Inertial" reference frames that are moving with constant velocity with respect to each other and to the fixed stars allow easy transformations of the description of motion among themselves. Forces (though not "paths") remain the same when viewed by observers in such frames. Newton's laws hold in these inertial frames. On the other hand, Einstein's general theory of relativity in principle allows us (or sometimes forces us) to declare the equivalence of gravitational fields and acceleration fields. As a practical matter it is frequently convenient to describe phenomena from a reference frame that is accelerating with respect to some other. For an observer *in* a rotating system, the easiest (and perfectly correct) description of events is usually in terms of centrifugal and Coriolis forces. Imagine the difficulty of explaining the cyclonic motions of the atmosphere on our rotating earth by using a reference system not attached to the earth, and so avoiding the use of the Coriolis effect. At any given point within a rotating system, these inertial forces should be considered equivalent to local gravitational fields. That is certainly the way a plumb bob considers them!

There are those, however, who avoid the scholasticism of talking about fictitious forces by preaching a greater heresy. The earth exerts a centripetal force on the moon, they say, and therefore the moon must pull outward on the earth. So far, who could deny them? But then the argument slips into the assertion that the latter force is centrifugal and that the moon is in equilibrium because in this action–reaction pair the centripetal force just balances the centrifugal. Oh, fallacy! The forces of interaction operate on *two* bodies, not *one*. The moon cannot suffer both an "action" and also its "reaction."

The action–reaction proponents were characterized long ago by Prof. William Fogg Osgood, Professor of Mathematics at Harvard, in his book, *Mechanics* (Macmillan, New York, 1937), p. 102: "Some of them are good citizens. They vote the ticket of the party that is responsible for the prosperity of the country; they belong to the only true church; they subscribe to the Red Cross drive–but they have no place in the temple of science . . ."

In this new year, observe how others are impelled toward the center to maintain their circular paths. For yourself, when traveling in circles, flee the center. Finally, regardless of what circles you travel in, resolve to use but one reference frame at a time.

January 1973

Circular Reasoning

A round the halls here at Stony Brook we have been bemused by a simple little physics problem. It's probably as old as Newton, but I had never seen it before. So many obvious, yet subtle concepts of physics are involved that I thought you, too, might be amused if not bemused.

Arrange a turntable on which a bob can slide radially. Fasten the bob to a radial spring with the other end of the spring fastened to the axis. For the sake of simplicity, assume that the spring has zero length when it is not stretched—a Slinky on a carousel would be a good approximation. With the table turning at constant ω, the bob is at a radius r. The spring provides the centripetal force kr, which must be equal to $m\omega^2 r$. Mirabile dictu! The action is independent of r. A rider on the table would observe that the restoring force of the spring exactly balances the centri*fugal* force experienced by the bob, provided that $k = m\omega^2$. Without exerting any radial force, and so without doing any work, the rider could move the bob to a larger radius. At the larger radius, of course, there is more energy stored in the spring ($\frac{1}{2}kr^2$) and the magnitude of the rotational energy [$\frac{1}{2}(mr^2)\omega^2$] is also larger. Something for nothing! Isn't that neat? We knew physics could do it. As an extra bonus, the bob obtains extra angular momentum ($mr^2\omega$). But where is the torque that produces this angular momentum? We're just sliding the bob out radially, with zero radial force, perhaps along a friction-free radial groove.

Mull it over awhile.

If still bemused, consider first the situation in the rotating reference frame. True, the radial force required to go to larger radius is zero, and the potential energy of the spring does increase. The *magnitude* of the rotational energy also increases, but consider its polarity. The rotational energy is zero when r is zero. As the bob moves to larger radii, it is falling in the centrifugal force field. In other words, its energy is becoming more negative; it is becoming more deeply bound. At all radii, $|1/2kr^2| = |1/2(mr^2)\omega^2|$, since $k = m\omega^2$. At $r = 0$, the total energy is zero. As the bob moves to larger radii, the positive potential energy of the spring increases and so does the magnitude of the negative rotational energy. Their sum, the total energy, remains zero.

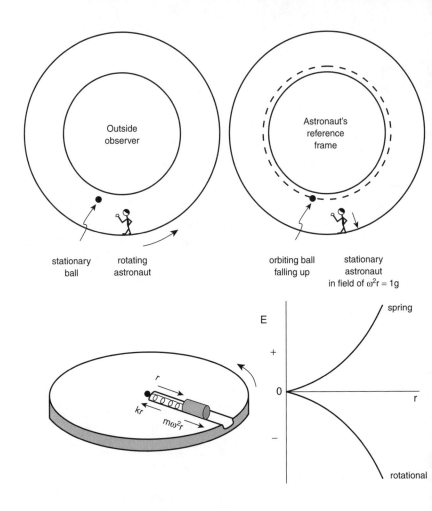

What about the angular momentum in the rotating reference frame? For an object constrained to move only radially, there is no angular momentum in a rotating reference frame, so there is no problem. There is, however, a Coriolis force. As the bob is moved out with velocity v_r, no radial force is required but there will be a Coriolis force $(2m\,\vec{v}_r \times \vec{\omega})$ directed opposite to the tangential velocity. If we slide the bob in a radial channel we'll never notice—but keep it in mind.

From an outside reference frame, looking down on the carousel, we have troubles. Both spring energy and rotation energy are positive—no doubt about it. They both get larger but the radial force needed is still zero. Furthermore, in this reference frame there *is* angular momentum, and it's increasing. There must be a torque, both to increase angular momentum and to provide the extra energy. The torque is provided by the walls of that radial channel, responding to the Coriolis force. For the bob to move outward, the walls must shove tangentially, thus exerting a force through a lever arm for the torque, and a force through a (tangential) distance to provide the necessary extra energy. The work is done by the turntable which must keep going at constant ω in spite of the countertorque.

If you liked that one you'll also like the one about the astronaut holding a baseball in a toroidal space colony which at first is at rest. The astronaut lets go of the baseball which just floats there. Now the giant torus (still free of air) is rotated until the centrifugal force field inside is 1 g. The astronaut has been holding on to the wall and can now stand there, with her feet outward radially, comfortably watching the ball go whizzing by. (An outside observer would say that the ball is still stationary.) How does the astronaut explain this situation? In the reference frame of the colony the "weight" of the ball is $m\omega^2 r$ and the direction is radially outward (down). The ball appears to be in orbit, but instead of continually falling downward, like the moon toward the earth, the ball is continually falling upward. The astronaut, whose name is Corey Olis, knows her way around and can explain all this quantitatively with circular reasoning. How about you?

April 1981

Misconceptions for Grown-Ups

We all know how hard it is to cure student misunderstandings. A mind convinced against its will is of the same conception still. Failing easy correction through instruction, the young may yet find truth through the aging process. As St. Paul is reported to have said, "When I was a child, I spoke as a child, I understood as a child, I thought as a child, but when I became a man, I put away childish things."

Then there are the misconceptions that we adults have. We know very well, for instance, that Feynman diagrams are not bubble-chamber tracks. There's a temptation, however, to view the diagram of the exchange of a virtual photon between two electrons as a picture of the actual event, with the wiggly line launching from one electron, thus changing its path, and colliding with the other electron to conserve momentum and change its path. Thus the exchange has produced repulsion between the two negative particles. This model is hard put to explain attraction between a negative and positive particle, but of course, such a view of the model is a misconception. The Feynman diagram is a shorthand for the integrals that must be formed and evaluated to find the probability of an interaction. The lines keep track of the different possible subinteractions that must be calculated. They aren't supposed to picture the paths and collisions of the particles themselves.

There's a similar problem that many of us have. It's the anthropomorphic fallacy that photons travel from place to place along routes, and that sometimes particles are waves—or at least behave like waves. We still think that in the double-slit experiment the photons go through one slit or the other, somehow knowing whether that other slit is open. It's particularly bothersome that the photon can figure this out even when there's only one photon at a time in the apparatus. Electrons can behave like that, too. It only helps a little to think of the electrons turning into waves. But, of course, the quantum mechanics never tells us the route these particles take. It only predicts (and always correctly) the probability of observable events. If we try to observe the route, we have a new experiment and can get a correct prediction of probabilities for the new event. Furthermore, it's the probability function that has the wave property. The particle isn't a wave and doesn't behave like a wave. That's as true of photons as it is of electrons.

One of the complicated equations with which we bedevil our students is due to Bernoulli. It is derived by applying the work-energy theorem to a slug of incompressible fluid that is being pushed with zero viscosity along streamlines. As von Neumann said, "It applies to dry water." Once having derived the equation, however, we like to apply it to all sorts of phenomena that don't satisfy the constraints. For instance, we ascribe lift on airfoils to the air going over the top of the foil with greater speed and thus less pressure. But air isn't incompressible and the streamlines are all mixed up and dissipated with turbulence. There are three terms to Bernoulli's equation, each apparently satisfying the dimensions of energy density. The first term is pressure of the fluid in a small region, the second is kinetic energy density of that fluid, and the third is the gravitational potential energy density of the fluid. It appears that three forms of energy density are present, and the first of them is pressure! One might be tempted to think that if a fluid flows into a narrower region so that it is flowing faster, the extra kinetic energy must be supplied by the stored compression energy measured by the pressure. How odd. Actually, the pressure term enters the derivation as the agent of some outside source (perhaps a pump) that is driving the fluid flow. This outside source does work on the fluid, which changes its kinetic and potential energy. When the work is done, it doesn't hang around in the form of compression energy, especially if the fluid is incompressible.

Even if the fluid is water, there's not enough compression energy under normal circumstances to have any effect on the fluid motion. Aside from that strange misconception, how would you apply Bernoulli's equation to the flow of water through a garden hose? Water is not dry.

Speaking of liquids, did you ever try to explain to students why the pressure in a glass of water is proportional to the depth, but the temperature is constant all the way down? To exert that increased pressure, aren't the little H_2O's hitting the walls harder and harder, dashing around with greater speed? Therefore, shouldn't the water be hotter at the bottom? Elementary-school texts foster this misconception by showing three diagrams of molecules in gas, liquid, and solid. The little circles in the gas diagram are far apart, but in the diagram of the solid they are shoulder-to-shoulder. Unfortunately, in the liquid they are also well separated and are pictured as dashing about. If that were so, the density of the solid would be much greater than that of the liquid, but for most materials, the solid and liquid phase have about the same density. Therefore, in the liquid phase, the individual molecules are also shoulder-to-shoulder and aren't dashing anywhere. In their neighborhood wells, they vibrate just like molecules in a solid, but by various means they slowly exchange places with their neighbors and meander around. Liquid pressure is not caused by translational bombardment, but is due to compression of the molecules against each other. The temperature, however, is related to the average kinetic energy of oscillation of the molecules in their temporary wells.

In the March issue there was a critique of analyzing rolling motion in terms of an axis that is "instantaneously at rest." The notion that there is such an axis is very popular. Let us bypass the question as to whether such an axis exists in a useful way (it doesn't), and pass on to the strange doctrine that if such an axis exists, the friction force created at that point must be due to static friction. (After all, that part of the wheel touching the road is motionless.) Rolling friction doesn't work that way at all. If it did, rolling friction would be greater than sliding friction, and we know that's not true. The energy lost due to rolling friction is expended in deforming the wheel and road. Traction may also be present, but this dissipates no energy unless there is slipping.

It's just barely possible that someone may want to disagree with me about some of these conceptions. Oh, good!

April 1998

On the Joy of Discovering My Ignorance

I t happened again. I was confronted with a simple, common physical phenomenon, and I didn't know the cause. Do you know why some icebergs, under overcast skies, are blue? "It's scattering," I said. Blue sky explanation. But why is the effect enhanced on overcast days and very faint when the Sun is bright? Furthermore, as I rotated my sunglasses, it was apparent that the blue light was not polarized.

When I returned from our Alaskan adventure, I consulted my fellow wizards at Stony Brook, who concluded after many blackboard discussions that they didn't know either. I could find no mention of blue icebergs in Minnaert or other books on atmospheric colors, nor did the author of a well-known text in optics know. Finally, I remembered a 1985 *TPT* article co-authored by Craig Bohren ("Colors of the Sky," 23, 267) and called him. He had described the phenomenon about a dozen years ago in the *Journal of the Optical Society of America* (73, 1646 [1983]). He has a good explanation—I think. Water is almost transparent in the visible, but not quite. It absorbs slightly in the red. However, you can see the remaining blue color only if there are discontinuities to reflect light from the interior. Deep holes in snow are blue inside. Icebergs must have bubbles or accumulation discontinuities to reflect back the unabsorbed blue from the interior. Are you satisfied? I think it's right, but there are a few things I want to try this winter. What a delightful, and pretty, problem. The blue icebergs are gorgeous.

This happens all the time. I keep finding new things I didn't know, or thought I knew but now am not so sure. Why do some things get hot in a microwave oven? Is it just induction heating of a material that is electrically conducting? Or is the microwave frequency tuned to the vibration or rotation resonance of the water molecule? Liquid water doesn't consist of little H_2O's. There are clusters of molecules binding and unbinding. Surely the binding would prevent any simple resonances. Besides, other materials, such as butter, get hot in a microwave. Surprisingly, dry ice cubes do not. I've had a lot of fun worrying about this problem. Probably somebody knows the answer in complete detail and will write a letter and tell me.

You know why conduction electrons in metals contribute only

slightly to the specific heat? As all the textbooks point out, the Pauli exclusion principle forces electron pairs to fill the energy levels up to the Fermi level of several electron-volts. Except for those few electrons at the top of the Fermi sea, the other electrons can't accept thermal bumps of the order of 1/25 eV, and so they lie dormant. Then how come *all* these electrons can take part in electrical conduction, and also why are electrons responsible for most of thermal conductivity in metals? I think I know now, but until a few years ago I hadn't even realized that there might be a problem.

Then there was the Coke-bottle problem. One Monday morning a colleague appeared at my office door, lip-sounding the fundamental of an empty Coca-Cola bottle. He had expected to get a note with a wavelength equal to four times the bottle length. But this tone did not at all match the simple formula for a tube closed at one end. Before long a fair share of this distinguished faculty were playing soda bottles of various sizes. It turns out that none of us remembered the Helmholtz oscillator, which any nineteenth-century physicist would have recognized immediately. The formula for the fundamental frequency of a Helmholtz oscillator involves the volume of the bottle, but is not concerned with the length of the closed container. The bottle is not a tube.

$$ f_0 = \frac{v}{2\pi} \sqrt{\frac{A}{\ell_e V}} $$

(see figure; $\ell_e = \ell + 0.8 \sqrt{A}$)

Our rediscovery of what Helmholtz knew long ago solved the Coke-bottle problem—or did it? It turns out that if you are sounding a plastic bottle and pinch the sides just slightly, the tone vanishes. Helmholtz doesn't say anything about the shape of the volume. Why should ruining the symmetry ruin the oscillation? You don't suppose that we are hearing a beat between an axial and a radial oscillation, and when you pinch the bottle you destroy the radial oscillation?

Wouldn't it be boring if physics consisted of a bunch of formulas to memorize, and after enough practice we could solve everything? For-

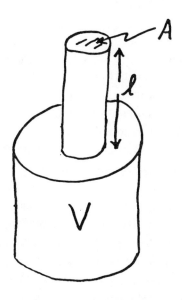

tunately, the list of things that I know that I don't know grows longer every year. Tomorrow, with any luck, I may discover something new that I don't really understand. It will probably be something that I have been teaching for years.

November 1994

The Plumb Line

│ n one of the visions of Amos, described twenty-seven hundred years
│ ago, Amos saw God standing by a wall with a plumb line in his hand.
│ Surely no simple instrument can have a grander lineage than that.
And the plumb line is indeed a simple instrument. Tie a weight to a
string and the weight will hang down with the string vertical—by definition. Once you know the direction of down, you know the direction of
up. It's hard to know the direction of up if you don't know the direction
of down.

To this day, people who build walls use a plumb line to make sure the
walls are vertical. It's also useful in the house when hanging wallpaper.

Physics students can use a plumb line as a pendulum. It's not quite a
simple pendulum, however. The bob is not a point, and the center of
percussion of the system does not quite coincide with the center of
mass. (The period is equal to

$$2\pi \sqrt{r/g},$$

where r is the radius to the center of percussion. You can find the center
of percussion by tapping the bob and locating the point where the bob
starts swinging without wobbling.) Of course, the period is this simple
only for small angles.

Newton did a lot of experimenting with pendulums because he was
worried about the apparent equivalence of gravitational mass and inertial mass. He compared the periods of pendulums that had bobs of
different materials. In the derivation of the pendulum period, mass
cancels out for the same reason that it cancels out of the formula for
acceleration due to free fall. If $m_{grav} = m_{inertial}$, then:

$$\not{m}_{grav}\, g = \not{m}_{inertial}\, a \tag{1}$$

and

$$\not{m}_{grav}\, g \sin\theta = \not{m}_{inertial}\, r\, \frac{d^2\theta}{dt^2}$$

If we really believe that plumb bobs always point down, we may be concerned about their behavior when used as conical pendulums. It appears that at any given instant in its circular motion, the bob is pointed partially toward the center of the Earth and partially *outward*. Evidently, by the plumb-line definition, the direction of down in this case is continually changing. Of course a rider on the bob would judge the situation differently. As the world swirls around the reference frame of the bob, the rider would claim that the local gravitational field is just the vector sum of the Earth's field and a constant centri*fugal* field. The rider going around the bend on the amusement park car would agree.

We needn't appeal to demon riders on rotating bobs or children in amusement parks. Airplane riders experience inertial fields. As Betty Wood pointed out years ago in her book, *Science for the Airplane Passenger*, you need not take your plumb line on board. Just hold an airline magazine lightly at one corner during takeoff or landing. You will see the diagonal line swing backward during takeoff as your local field becomes the sum of gravity toward the center of the Earth and the *backward* inertial field. If you happen to have a helium balloon, you can observe the way an anti-plumb-bob works. Trust it. It always points *up*. You can try this in a car, too, while accelerating, braking, or going around a corner. (Let a passenger hold the balloon.)

An astronaut in orbit might be dismayed to find that plumb lines don't work at all. Evidently there is no down up there. If the astronaut looks out and sees the Earth (below?) and believes in Newton's universal gravitation, then the astronaut might properly claim that she is also in a centri*fugal* field that just cancels Earth's gravitational field.

At this holiday time it should bemuse us to realize that such a primitive and simple device behaves in ways that are profound. If we take one step beyond any topic in introductory physics, we are in the midst of grand revelations or current research or mystery or all three. The simple behavior of the plumb line leads us to the subject of reference frames and the foundations of the general theory of relativity. It's a wise student who knows which way is truly down.

December 1994

Justifying Atoms

A student-teacher phoned me the other day. His supervising teacher had assigned him the task of persuading a ninth-grade class that atoms exist. Now, everybody knows that atoms exist. You learn it in third grade. In all succeeding grades the existence of atoms is just assumed. But can you prove it to ninth graders?

How would you answer? Of course you could appeal to textbooks or encyclopedia references. Those sources say that atoms exist. Or you could describe Rutherford scattering experiments that reveal the nuclear atom. You could also cite x-ray scattering that shows the arrays of atoms in crystals, or explain the new tunneling microscopes that trace the profiles of atoms on a surface.

An appeal to authority should satisfy ninth graders ("I say it's true, the textbook says it's true, it will be on the test"). They probably aren't old enough to appreciate the geometry of x-ray scattering patterns, and Rutherford's arguments about the nucleus work best after atomicity itself has been accepted. Still, atoms are sort of our stock-in-trade. Don't we have any hands-on experiments that make atoms tangible?

Probably you have a good way of answering this question. The best I can come up with goes like this. Due to an historical quirk in our educational system, most children study chemistry before they study physics. Therefore, they know about making and separating chemical compounds. In the process they have to learn about the laws of definite and multiple proportions. Now in the case of water, the law of *definite* proportions says that no matter where you get the water, when you dissociate it into hydrogen and oxygen, you will get a ratio of 8 between the mass of oxygen gas and the mass of hydrogen gas. You will also get twice the volume of hydrogen compared with oxygen. That's interesting but doesn't prove anything about atoms. Other compounds also yield definite proportions of mass and volume when they are broken down, but the ratios are not necessarily small integers. However, the law of *multiple* proportions is more remarkable. If you dissociate hydrogen peroxide, the ratio of oxygen mass to hydrogen mass is 16. That's exactly twice the ratio for water. It seems plausible to represent the compounds as H_2O and H_2O_2, representing the combination of chunks of matter. Of course, ninth graders aren't going to take apart the oxides of water, and compare masses. They can, however, do a simple electrolysis experiment with water and observe the volume ratios of the two gases produced. The gases can even be identified with reasonably safe tests using a glowing splinter. Perhaps then the students would be willing to accept textbook results for the mass ratios of the two compounds.

There's a good experimental analogy for the law of multiple proportions. It was much used at the introductory level in the "alphabet" courses of the 1960s. Have the students measure mass ratios for compounds of steel bolts and nuts. A bolt can correspond to oxygen and a nut to hydrogen. You can thus create H_2O with a bolt and two nuts, or even create a fairly stable H_2O_2 by using one nut to fasten together two bolts, one of which carries an extra nut. The mass ratios can thus demonstrate the laws of definite and multiple proportions and illustrate the difference between them.

Atomicity was not generally accepted until the beginning of the twentieth century. One of the most prominent European chemists, Ostwald, could explain the law of multiple proportions without using atoms. In the foreword of the second volume of his definitive book on kinetic theory, Boltzmann in 1898 wrote that one reason for demonstrating the power of the theory was that the idea of atomicity seemed to be losing ground. The clincher in favor of atoms turned out to be the Brownian motion that had first been observed in 1827. In one of the four great papers that Einstein wrote in 1905, he worked out the quantitative details for the distribution of Brownian particle migration. A few years later, Jean Perrin experimentally confirmed Einstein's predictions, which had been based on the kinetic theory of matter. In the process Perrin obtained a good measurement of Avogadro's number and won the Nobel prize in 1926.

With the right apparatus it's easy to see Brownian motion. You can buy a small chamber made for the purpose, with windows on the side and a hole and rubber snifter on the top. Shine a microscope lamp in one window and look in another at right angles, using a 50× microscope. Sniff some smoke into the chamber. The smoke particles will look like tiny stars. After the swirling motion has died down, the particles will float at low speeds, but they will also dance and migrate in a random walk. They are being bombarded on all sides by the air molecules, which of course are much smaller. Nevertheless, the smoke particles share in the molecular kinetic energy: $1/2mv^2 = 3/2kT$. The average speed comes out to be about 0.1 mm/s. The speed is small but clearly large enough to explain the jiggling motion seen in the microscope. For a smoke particle with a diameter of 10 μm, the volume (and mass) is about 10^{13} times that of an air molecule. The situation is analogous to the *Queen Mary* being buffeted on all sides by a sea of corks. The astonishing feature is not that the corks can jostle the ocean liner, but that the *fluctuations* in bombardment can move the ship. The phenomenon is even more dramatic with the micron droplets in the Millikan oil-drop apparatus. Brownian motion is about as close as we can come to "seeing" atoms and proving their existence, but to be properly impressed you must have some quantitative feel for what you are seeing.

On second thought, most ninth graders are pretty young to understand the experiments and arguments that demonstrate atomicity. Maybe it's like calculus and other facts of life that we should keep from them until they are old enough. That ought to intrigue them.

November 1997

Seeing Atoms

The fourth-grade teacher told her class that scientists cannot yet see atoms. My granddaughter, Marin, raised her hand and said, "My grandaddy can see atoms." "No, no, you must be mistaken," the teacher answered. "Our text says that atoms are too small to be seen." "Well," said Marin, "my grandaddy takes pictures of them." She called me that evening. "Please send pictures."

Can we really see atoms? Well, of course we can. We've been seeing them for years. The only thing that needs a little explanation is what we mean by "seeing." Now let's see. Do you watch television? Surely on the screen you see people and objects, moving and in living color. Actually, the things you think you see are far away, and perhaps distant in time. In some cases the things you see never existed except in the form of regions of magnetized film. Have you seen weather patterns on the news, including the views of cloud cover taken by a satellite? All those pixels on your screen, which you see, are actuated by electrical signals directed by a very complicated network of devices. Talk about radical constructivism! The original scanning may have been done by beams of several different electromagnetic frequencies, not necessarily in the visible.

A small fraction of the light leaving the TV screen may enter our eyes. There's where the real reconstruction and analysis begins. The light doesn't just pass through our eye lens, its path bending appropriately at the boundaries. The only way the light can pass through material is to interact with the molecules of the material. The photons are absorbed and re-emitted—or, if you prefer, the electromagnetic waves set the local oscillators into vibration and they reradiate. The light rays make their Snell turns because all other paths lead to wave cancellation. But understanding the optics is child's play compared with knowing what happens when the coded nerve signals march into the brain. Apparently there is some system up there on my shoulders that can match the signals to memorized patterns. Aha! I recognize that sequence of pulses. I have seen a human on the TV. Of course I have.

You don't have to start with electromagnetic pulses. Send a beam of high-frequency sound waves into a human body. A microphone picks up the attenuated signals (are you listening or seeing?), converts them into electrical signals that control the intensity of an electron beam in a display tube—and there! I have seen a fetus.

All the stars in the heavens, except for our Sun and a few others, are point sources. No matter how big our lenses or mirrors, we do not see them as disks. They are points. And yet, we all know how much we can see by looking at the stars. We send their light through spectrometers and so see the elements in the star's atmosphere. We see that its spectral lines are bluer or redder than the ones in our laboratory, and so we see that the star is coming toward us or away from us. We see how fast it is rotating, how hot it is, and how old it is.

About once a year, in recent years, the newsmagazines carry a story that scientists have just been able to see atoms. They show pictures of arrays of blobs. There is nothing really new about this seeing. It isn't as if the "scientist" (white-coated, distinguished, lean, certainly male) has been able to peer down through a new type of microscope and finally has "seen" such little things. As always, the probe that sees must be smaller than the object being seen. There are then various devices that measure the scattering of these probes and reconstruct the geometry of the scattering object. The advances of the last few years have really been in the power and sophistication of the computer-like devices that receive, catalog, and then display the scattering in graphical forms that we humans can interpret.

Rutherford saw the nucleus of the atom in 1913. His probe was the alpha particle, his recognition pattern was the $\sin^{-4}(\theta/2)$ graph of scattered intensity. In the winter I see rabbits, raccoons, and foxes in my yard at night. They leave tracks in the snow that I have come to recognize. Years ago I took cloud-chamber pictures of electrons and protons and deuterons and alpha particles. I recognized them by their tracks.

If you protest and say, "You didn't really see the particles or the raccoons. You only saw their tracks," then I must confess that I really don't see you. To be sure, the process of seeing a particle or seeing you is very similar. Visible light scatters off your face or scatters off the cloud-chamber droplets. A small fraction of that light enters my eyes, goes through complicated interactions to my brain and then, in a process that is not well understood, matches recognition patterns. Have I seen the particle? Have I seen you? The particle has only a few attributes—seven or so qualities such as mass, charge, and spin. I can measure each and all of them. You, however, have many attributes. All I know of you has to be interpreted from a few superficial scattering patterns. It would be presumptuous of me to say that I see you. The particles are real and I see them completely. You are, at best, a hypothetical construct.

Of course, I didn't tell Marin that. I just sent her some chamber pictures and talked about raccoon tracks in the snow. She knows I can see atoms, and someday, so will she.

January 1992

Believing Is Seeing

Here's a picture that makes manifest the reality of subatomic particles. The picture is filled with interesting evidence. The main beam of sparse, gently curved tracks is caused by 870-MeV negative pi mesons passing through liquid hydrogen in a bubble chamber. The paths were curved because there was a perpendicular magnetic field in the region. Every so often a pion knocked off an electron from one of the hydrogen atoms, producing a small spiraling track. There are two places in the chamber where the pions struck hydrogen nuclei. The one near the top was an elastic collision, with the pion glancing off to the right, knocking the proton to the left.

The most surprising event starts in the center about one-fourth of the way up from the bottom. A negative pi meson collided with a positive proton yielding two neutral particles, a lambda zero (Λ^0) and a theta zero (θ^0). (The θ^0 is now known as a K^0.) The θ^0 is a heavy meson, composed of a quark and an antiquark, one of which has a unit of positive strangeness. The Λ^0 is a baryon, a combination of three quarks, one of which has a unit of negative strangeness.

Now comes the special feature. We know that the pion must have hit a proton and turned into at least one neutral particle because the pion track just stops. A high-energy particle can't just stop, but if neutral particles were produced they would leave no tracks. Therefore we look downstream for evidence of the resulting neutral particle or particles. Sure enough, a few centimeters away (in the original chamber), the Λ^0 decayed into a positive proton and a negative pi meson. You can determine the signs of their charges by comparing their curvature in the magnetic field with that of the negative pions in the beam. Furthermore, the more massive proton is traveling with a smaller speed and leaves a more dense track. If the Λ^0 had been traveling close to the speed of light, it would have taken 1×10^{-10}s to travel 3 cm. Meanwhile the θ^0 must have traveled off to the left, and its decay into two pions should yield another V-shaped pair of tracks. That's what is usually seen in these associated events. Instead, off to the left, we see a single track with high density and slight curvature in the direction required by a positive charge. It's a proton! But the θ^0 can't decay into a proton. The θ^0 must

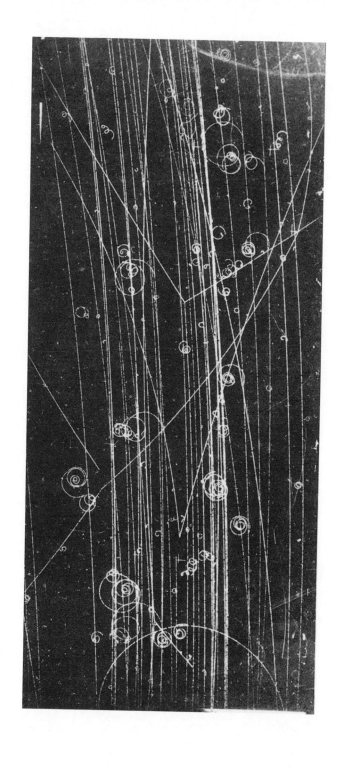

have hit the proton and knocked it forward like a billiard ball collision. If so, the θ^0 must have lost a lot of energy and bounced *backward*. Sure enough, we see the V tracks of the two decay pions, but they received very little forward momentum from their parent θ^0 and so go off almost opposite to each other, their kinetic energy provided by the difference in mass between the θ^0 and the two pions.

How can anyone see a picture like this and not believe that the subatomic particles are little real objects?

January 1992

The Season of Harmonics

leighbells ringing, dulcimers twinging. The longest nights are here and visions of Fourier series dance through our heads. All those lovely holiday sounds can be produced by combining sinusoidal waves that have frequencies that are integral multiples of some fundamental. To be sure, the bells aren't generating harmonics of their fundamentals, but the strings are, and on paper, at least, we can synthesize any repetitive function with a Fourier series.

Each instrument has a characteristic combination of overtones for each note, though the recipe may change from the first attack of the note to the continuing sound. When a horn begins a note, it trumpets out brilliant overtones. A struck piano string vibrates at first with many frequencies. If the note is held, the higher harmonics die away, leaving a single sinusoidal wave without character or timbre. As a rule of thumb, faithful reproduction of those characteristic beginning sounds requires at least ten harmonics, and twenty is better.

Consider the function shown at the top of our holiday centerfold. It is simply a sine function, $f(t) = b \sin 2\pi f_0 t$. If this represented a musical note, we would hear only the fundamental with a frequency f_0. It is the sound a tuning fork would make, or a string after all the harmonics have died away.

On the next line of the centerfold we show a more complex function made up of a fundamental and its first harmonic.

$$f(t) = b_1 \sin 2\pi f_0 t + b_2 \sin 2\pi \, 2f_0 t$$

In this case, $b_2 = 1/2b_1$. If the fundamental frequency is 440 Hz (the A that the orchestra tunes to), then the first harmonic is an octave higher at 880 Hz. Notice that in the formula and the graph these are both sine waves starting in phase with each other. If the phase difference between the two frequencies changes, the pattern seen on an oscilloscope would be different, but we would hear the same chord.

The general Fourier series for a function periodic in $T = \dfrac{1}{f_0}$ is:

$$f(t) \equiv \tfrac{1}{2}a_0 + \sum_{n=1}^{\infty} (a_n \cos n\omega_0 t + b_n \sin n\omega_0 t), \text{ where } \omega_0 = 2\pi f_0.$$

What's missing in this recipe is the relative amplitudes of the coefficients, a_n and b_n. How much of each harmonic do we mix in? About two centuries ago, Jean Fourier showed how to calculate the values of the coefficients. The method depends on the fact that

$$\int_{-\pi}^{\pi} \sin n\omega t \sin m\omega t \; dt = 0 \quad n \neq m$$

$$= \pi \quad m = n.$$

Similar equations hold for the cosine function. This property of the sinusoidal functions is called *orthogonality*. The nth coefficient can be evaluated by multiplying both sides of the function equation by the appropriate sine or cosine term with frequency $n\omega_0$ and integrating over one period. For instance, to find a_1:

$$\int_{-\pi}^{\pi} f(t) \cos \omega_0 t \; dt = \tfrac{1}{2} a_0 \int_{-\pi}^{\pi} \cos \omega_0 t \; dt +$$

$$a_1 \int_{-\pi}^{\pi} \cos^2 \omega_0 t \; dt + a_2 \int_{-\pi}^{\pi} \cos \omega_0 t \cos 2\omega_0 t \; dt +$$

$$\ldots + b_1 \int_{-\pi}^{\pi} \cos \omega_0 t \sin \omega_0 t \; dt + b_2 \int_{-\pi}^{\pi} \cos \omega_0 t \sin 2\omega_0 t \; dt + \ldots$$

$$= 0 + a_1\pi + 0 + \ldots + 0 + 0 + \ldots$$

Therefore, $a_1 = \dfrac{1}{\pi} \displaystyle\int_{-\pi}^{\pi} f(t) \cos \omega_0 t \; dt.$

Similarly, multiplying through by $\cos n\omega_0 t$ yields:

$$a_n = \dfrac{1}{\pi} \int_{-\pi}^{\pi} f(t) \cos n\omega_0 t \; dt.$$

Multiplying through by $\sin n\omega_0 t$ yields:

$$b_n = \frac{1}{\pi} \int\limits_{-\pi}^{\pi} f(t) \sin n\omega_0 t \, dt.$$

The third line of our centerfold (see pp. 316 and 317) shows the wave pattern for a trumpet note. The Fourier recipe calls for almost equal amplitudes for all the harmonics. Notice the way the high frequencies sharpen the vibration, giving the trumpet its brilliant sound.

On the next to the bottom line of the centerfold we show the Fourier buildup of a square wave. It's hard to generate a square wave, either electrically or with audio devices. The fast rise and sharp corners demand a whole range of high harmonics. We used only the fundamental and the next two harmonics for a pretty good approximation. Overlapping the square wave in the diagram is the curve of the fundamental, then the sum of the fundamental and first harmonic, and then the sum of the fundamental, first, and second harmonic.

We have arranged the square wave, shown in the following figure, to have values convenient for calculation and to look more like a sine wave than a cosine. Therefore, the function is odd; that is, $f(t) = -f(-t)$. All of the a_n will be zero, since $\cos n\omega_0 t$ is an even function.

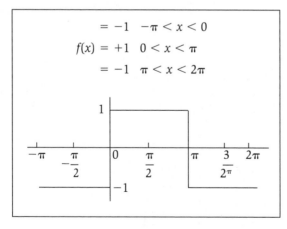

$$f(x) \begin{array}{ll} = -1 & -\pi < x < 0 \\ = +1 & 0 < x < \pi \\ = -1 & \pi < x < 2\pi \end{array}$$

$$b_n = \frac{1}{\pi} \left[- \int\limits_{-\pi}^{0} \sin n\omega_0 t \, dt + \int\limits_{0}^{\pi} \sin n\omega_0 \, dt \right] =$$

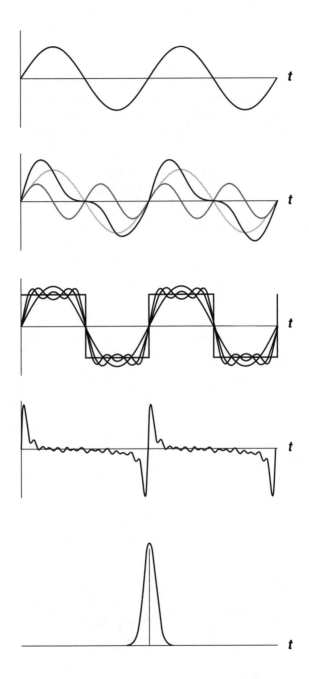

Harmonics

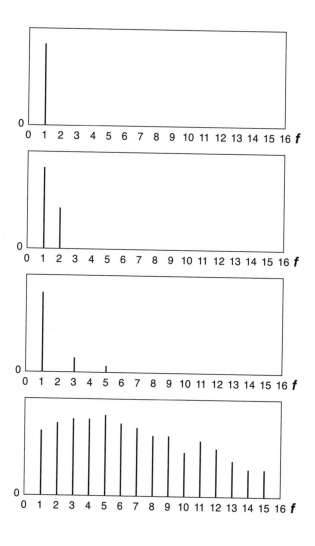

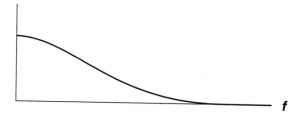

$$\frac{1}{n\pi}\left[\cos n\omega_0 t\Big|_{-\pi}^{0} - \cos n\omega_0 t\Big|_{0}^{\pi}\right]$$

$$= \frac{1}{n}\pi\,[2+2]\ \text{for odd } n, \quad \text{and} \quad = \frac{1}{n}\pi\,[0+0]\ \text{for even } n$$

Therefore, the series for this square wave is:

$$f(t) = \frac{4}{\pi}\sin \omega_0 t + \frac{4}{3\pi}\sin 3\omega_0 t + \frac{4}{5\pi}\sin 5\omega_0 t + \ldots$$

On the right-hand side of the poster we show the recipes for the amplitudes of the harmonics. Notice that the original patterns show amplitude as a function of time. The right-hand graphs show amplitudes of the partials as a function of *frequency*. This other way of representing the original information is called a Fourier transform. The term is usually applied to transformations of nonperiodic functions.

The bottom curve in the centerfold shows a single Gaussian pulse, $f(t) = \exp(-t^2/I^2)$. Because the pulse is not periodic, its Fourier transform must be the *integral* of a continuous function of the frequencies instead of a *sum* of integral multiples of a fundamental. The transform equation is

$$f(t) = \frac{1}{\sqrt{2\pi}}\int_{-\infty}^{+\infty} F(\omega)e^{-i\omega t}\,d\omega.$$

The function $F(\omega)$ is called the Fourier transform of $f(t)$. Knowing $F(\omega)$ is equivalent to knowing $f(t)$ since $F(\omega)$ is a recipe of the spectral distribution necessary to represent the function $f(t)$. In the case of our Gaussian pulse, the Fourier transform is

$$F(\omega) = \frac{T}{\sqrt{2}}e^{-T^2\omega^2/4}.$$

Adding all those cosine terms together, modulated by $F(\omega)$, creates the pulse maximum at $t = 0$, where the terms are all in phase. At later or earlier times the cosines (with steadily increasing frequency) get increasingly out of phase and cancel each other, leaving only the central pulse. The narrower the original pulse, the greater the spread of frequencies that it contains, and thus the wider the pulse on the transform graph of amplitude as a function of frequency. For instance, if the half-width of our Gaussian pulse is 1 s, then the width of its transform is 4 s^{-1}. If the pulse is narrowed to 0.1 s, a greater spread of frequencies is needed to create it and the transform width increases to 40 s^{-1}.

These are all good things to know as we enter the holiday season. Surely your pleasures will be enhanced by Fourier analyzing the trilling of the flutes and the tintinnabulation of the bells.

December 1993

Almost everything twangs. Nuclei, atoms, molecules, crystals, bells, violin strings, pot covers, electrical circuits, bridges, yea the great globe itself. Disturb any one of these and it will oscillate around some equilibrium position, gradually dissipating the energy that disturbed it. If you drive the system at its resonant frequency, the oscillations may grow to alarming size. The duration of the vibrations of a disturbed system, and the sharpness of its response to a driving force, can be characterized by a dimensionless constant, Q.

Oscillations are the most common form of motion because they are caused by one of the simplest force laws: $F = -kx$, where F is the restoring force on an object that is displaced a distance, x, from its equilibrium position. To first approximation, for small displacements, Hooke's law does indeed describe the response of most bound systems. This is true of the displacement of atoms bound in their crystalline or molecular positions, and so is true on the macroscale of solids. A Hooke's law force generates sinusoidal oscillations.

Oscillations die down. The energy is dissipated by friction or radiation. If we assume that the friction is proportional to the velocity, which is a fair approximation for many cases, then the friction force term is

$$bv = b\frac{dx}{dt},$$

where b is the friction proportionality constant. The oscillator equation becomes

$$m\frac{d^2x}{dt^2} = -b\frac{dx}{dt} - kx,$$

which yields the familiar damped oscillation curve shown in our centerfold.

For ease of analysis, many texts (as well as common practice) define the following quantities:

$$\omega_0 = \sqrt{k/m} \quad \text{the "natural" resonant frequency}$$

$$\gamma = b/2m \quad \text{the damping factor, with dimensions of frequency}$$

$$Q = \frac{\omega_0}{2\gamma} = \frac{\sqrt{mk}}{b} = \frac{\omega_0 m}{b} \quad \text{the "quality" factor, which is dimensionless.}$$

Let's investigate the nature of the Q of a system. Evidently, the larger the friction proportionality constant, b, the lower the Q and the faster the oscillations die down. For very large Q, the damping is small and $\omega \approx \omega_0$. Even if $Q = 1$, $\omega = 0.87\,\omega_0$, which is low by only 13 percent. You can see the difference between oscillations with large Q and small Q in the diagrams on the centerfold.

For a Q of 600, about one percent of the energy is lost in each cycle. This is typical of a piano or violin string that sings for a second or so after it has been plucked. With a fundamental frequency of several hundred, the Q of such a string must be of the order of 10^3. A pot cover in our kitchen rings for about ten seconds until the intensity is down to about $1/3$. Its frequency is 660 Hz. Since it takes about 6,600 oscillations to reduce its energy by e, its Q must be 40,000! Seismic vibrations lose intensity very slowly, considering that it's the Earth that's vibrating, and have Q values of several hundred. An atomic transition producing visible light has a duration ($1/e$ time) of about 10^{-8} s. The period of visible light is about 10^{-15} s, and so the Q must be about 10^8. A gamma ray from the nucleus of Fe^{57} (when bound in a crystal—the Mossbauer effect) has a Q of over 10^{12}. Relatively, it rings forever. Note that a signal source with high Q generates a very long duration sine wave and therefore produces a very monochromatic signal.

Suppose now that a periodic external force with frequency ω drives an oscillator with natural frequency ω_0. If the driving frequency is close to the natural frequency, oscillations of large amplitude can build up. This phenomenon is important in both natural and human-made devices. The common behavior of any such system is characterized in terms of the ratio of driving frequency to natural frequency, ω/ω_0. When ω is much smaller or much larger than ω_0, the amplitude of the driven system is small. When ω is close to or equal to ω_0, the amplitude becomes large. The smaller the internal resistance or friction of the system (the larger the Q), the greater will be the resonant response when $\omega \approx \omega_0$ and the faster the amplitude will fall off as ω differs from ω_0. This behavior is shown in the graphs of our centerfold.

It is possible to demonstrate the qualitative effects of a driven resonant system with a very simple apparatus. Hang two pendulums from a stout, horizontal string whose tautness can be adjusted. If the two bobs have the same mass, the system will display complete interchange of energy between the two pendulums and the existence of normal modes. For demonstrating driven oscillation, however, make one of the bobs at least ten times heavier than the other. Then the heavy one will be the driver, and you can vary ω by changing the length of its string. Both amplitude and phase of response as a function of ω/ω_0 are easily demonstrated and observed.

The resonance curves of a driven system can be characterized in terms of the ratio of the frequency at maximum amplitude to the full width of the response curve. That ratio is equal to Q.

$$Q = \frac{\omega_{resonance}}{2\Delta\omega}$$

where $2\Delta\omega$ is the full width of the *amplitude* curve at the height, where $A = 1/\sqrt{2}\, A_{resonance}$, or, where $2\Delta\omega$ is the full width of the *energy* curve where $E = \frac{1}{2}\, E_{resonance}$.

For instance, a typical Q value for a crystal model radio is about 10^2. If the receiver is tuned to 1,000 kHz (the center of the AM band), the response curve would have a full-width (at half-maximum) of 10,000 Hz. This band width is necessary in order to carry the range of audio frequencies covered, which for AM radio is only 5,000 Hz.

Systems with high Q have low internal resistance, their ringing decays slowly, and their response to an external driving force is sharp and strong at the resonant frequency. The oscillations of a low Q system decay rapidly and the system responds sluggishly to a driving force, even at resonance. These criteria apply only to physical systems, of course, but at this midyear time physics teachers can't help but wonder how to characterize their current students. If you disturb their mental equilibrium, do you detect a ringing for a while? Have you found that their enthusiasm grows dramatically as you approach their natural interest frequency, which we hope is on the subject of physics? For the New Year we wish that all of you may have classes with resonant frequencies to match your own and students of the very highest Q.

$$F = ma = m \frac{d^2x}{dt^2} = -kx$$

yields

$$x = A \sin (\sqrt{k/m}\, t + a) = A \sin (\omega_0 t + a),$$

where ω_0 is the angular frequency, $\sqrt{k/m}$, and a is the starting phase. The larger the spring constant, k, the higher the resonant frequency. The larger the mass of the object, m, the smaller the frequency, ω_0. The frequency of the system in hertz is $f_0 = \omega_0/2\pi$, and the period is $T = 1/f_0$. (Note that the differential equation requires that the second derivative of the distortion, x, with respect to time, t, is proportional to *minus* the distortion. The sinusoidal function has this property. The derivative of sine is cosine; taking the derivative of cosine yields negative sine.)

A damped oscillation is described by

$$x = \left(x_0 e^{-(b/2m)t}\right) \sin \left(\sqrt{\frac{k}{m} - \frac{b^2}{4m^2}}\, t + a\right).$$

The *amplitude* of the sinusoidal term is $(x_0 e^{-(b/2m)t})$. It is equal to x_0 at $t = 0$, but then decays to $1/e$ of its starting amplitude at $t = (2m/b)$. As $t \to \infty$, $x \to 0$. Without friction, $\omega = \sqrt{k/m}$. Note that the presence of friction decreases the frequency.

In terms of γ and Q, the actual resonant frequency is related to the "natural" frequency by

$$\omega^2 = \omega_0^2 - \gamma^2 = \omega_0^2 \left(1 - \frac{1}{4Q^2}\right).$$

The equation for damped harmonic oscillation becomes

$$\frac{d^2x}{dt^2} + 2\gamma \frac{dx}{dt} + \omega_0^2 x = 0$$

or

$$\frac{d^2x}{dt^2} + \frac{\omega_0}{Q}\frac{dx}{dt} + \omega_0^2 x = 0.$$

The solution is

$$x = x_0 e^{-\gamma t} \sin(\omega t + a) = x_0 e^{-\frac{\omega_0}{2Q}t} \sin\left(\omega_0 \sqrt{1 - \frac{1}{4Q^2}}\, t + a\right).$$

Another way to see the significance of Q is to calculate the amplitude of energy loss *per cycle* in a damped oscillation. Instead of expressing t in seconds, express time in terms of number of periods of oscillation, n:

$$t = nT_0 = n\,\frac{2\pi}{\omega_0}.$$

In these terms, $x = x_0 e^{-n\pi/Q} \sin(\omega t + a)$. The amplitude falls by e (2.7) in Q/π cycles, and the energy falls by e in $Q/2\pi$ cycles (since the energy is proportional to the square of the amplitude). In *one* cycle:

$$\frac{\Delta x}{x} = \frac{\pi}{Q}$$

and

$$\frac{\Delta E}{E} = \frac{2\pi}{Q}.$$

(Since $e^{-n\pi/Q} \approx 1 - n\pi/Q$, then, $x_0 = x_0$, $x_1 \approx x_0 - \pi/Q$, and $x_2 \approx x_0 - 2\pi/Q \approx x_1 - \pi/Q$, etc.) We can get rid of the 2π factor by observing that the energy falls by e in Q *radians* and that the fractional energy lost *per radian* is

$$\frac{\Delta E}{E} = \frac{1}{Q}.$$

(For instance, if $Q = 12$, the energy decreases by a factor of 2.7 in 2 cycles or in 12 radians. The fractional loss of energy is $1/2$ in one cycle or $1/12$ in one radian.)

If a system is being driven by a force, $F \sin \omega t$, then the magnitude of its amplitude (after any transients have died out) is

$$|A| = \frac{F/m}{\sqrt{(\omega_0^2 - \omega^2)^2 + 4\gamma^2\omega^2}} = \frac{F}{k}\frac{\omega_0/\omega}{\sqrt{\left(\frac{\omega_0}{\omega} - \frac{\omega}{\omega_0}\right)^2 + \frac{1}{Q^2}}}.$$

When $\omega = \omega_0$, the amplitude becomes $Q F/k$. The Q factor acts like an amplifying term. In an electrical LCR series circuit driven by a signal with voltage amplitude V, the voltage across the capacitor becomes (Q_{value}) V at resonance, an amplitude that may well destroy the capacitor.

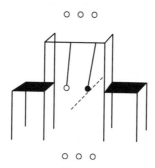

To transform these formulas to describe LRC electrical circuits, substitute L for m, R for b, and $1/C$ for k. The damping factor becomes $\gamma = R/2L$ and the quality factor becomes $Q = \omega_0 L/R$.

December 1990

Dimensionless Constants

As you know, our goal for this journal is that a physics teacher will find every article, note, column, and cartoon useful, preferably instructive, and at least of casual interest. Therefore we are particularly pleased with the exchange of letters about fluid flow carried in the Notes section of *The Physics Teacher*, September 1982. In that exchange a reasonable and instructive demurral about an earlier article has evoked a detailed and useful expansion of the original material.

The item in question is the significance of the Reynold's number, the best known of the dimensionless constants of hydrodynamics. Actually, we rarely deal with Reynold's numbers in introductory physics. The treatment of dimensionless constants is usually left to advanced courses in physics and engineering. In introductory physics our students have enough trouble dealing with relationships between two variables, for example, $v(t)$. It takes older heads to worry about phenomena involving many variables. One of the simplifying techniques for this task is to group variables into dimensionless combinations, producing parameters that characterize regimes of behavior.

The physical significance (or indeed, the dimensionality) of Reynold's number is not clear from its final formula $R = Lv\rho/\eta$. An object with "characteristic length," L, moves with velocity, v, through a fluid with density, ρ, and viscosity η. In the derivation, Reynold's number starts out being defined as the ratio of the inertial reaction of the fluid being shoved aside by the object, to the viscous drag experienced by the object.

If R is much less than 1, we can ignore the inertial reaction. Viscosity dominates and the drag is proportional to v. If R is greater than 10^3, the viscosity force is unimportant (although it may still affect boundary conditions that in turn control onset of turbulence). In this regime, drag is proportional to v^2. The exact boundary regions for R depend on the shape and surface texture of the object.

There is another vital use of the Reynold's number besides specifying drag dependence on velocity. The dimensionless constants are guides to scaling phenomena from one magnitude to another. For instance, a child's model airplane does not at all have the flight characteristics of its large prototype. The Reynold's number for the model is vastly different from that of the real airplane. The wingspread, L, is smaller, and so is the velocity. Consequently, wind-tunnel demonstrations in school science fairs are usually meaningless.

An important dimensionless constant in the atomic world is the fine structure constant. In many older texts and handbooks it is written as $\alpha = e^2/\hbar c$, where c is the electron charge in electrostatic units, $\hbar = h/2\eta$, where h is Planck's constant in erg seconds, and c is the speed of light in cm/s. In S.1., $a = (1/4\pi e_a)e^2/\hbar c = 1/2\mu_0 e^2 c/\hbar$, where e_0 is the permittivity of free space, $\mu_0 = 4\pi \times 10^{-7}$ is the permeability of free space, e is the electron charge in coulombs, c is the speed of light in m/s, and $\hbar = h/2\pi$ in joule seconds. In any consistent set of units, $a = 1/137$.

The fine structure constant got its name and first appeared in Sommerfeld's relativistic treatment of the Bohr hydrogen atom. The energy of a particular level with quantum number η is given by a principal term inversely proportional to η^2 and a correction term smaller by about a factor of $a^2 j$. Since a^2 is only 5×10^{-5}, the resulting fine structure in the spectrum of hydrogen is very fine indeed and hard to resolve.

It is not immediately obvious from the formula why this combination of electron charge, Planck's constant, and speed of light should be dimensionless and have anything to do with atomic spectra. The formula actually starts out as the ratio of the speed of the electron in the $\eta = 1$ orbit of the Bohr hydrogen atom to the speed of light.

$$\frac{h/mr}{c}$$

Or, a^2 is the ratio of the binding energy of the electron in the $\eta = 1$ Bohr orbit to the rest mass energy of the electron.

$$\frac{e^2/r}{mc^2} \qquad \frac{1}{4\pi e_0}$$

In any, or all cases, the ratio is a measure of the strength of the electromagnetic interaction. Because a has a value of only about 1%, it is possible to calculate quantum electrodynamic effects to great precision in a series of successive terms in a^2.

Physics teaching is another complex business involving many variables. Wouldn't it be interesting to combine these variables into a dimensionless quantity that rated good teaching? Perhaps we could use the quantity to determine suitability for teaching, or for salary increases. Let's make a list: (A) knowledge of subject, (B) knowledge of learning processes, (C) skill in teaching techniques, (D) importance to the teacher of financial rewards, (E) importance of freedom from administrative regulation, (F) importance of community prestige, etc. Perhaps our P.T. number could be the ratio ABC/DEF. Rate yourself for each variable on a scale of 1 to 10, and then compute your P.T. value. It could range anywhere from 10^{-3} to 10^3. If your score is over 10 you may be a good physics teacher but you're probably hurting financially. Of course, there's one dimension we didn't account for. That's enthusiasm. Enthusiasm for the subject, and enthusiasm for teaching it. That quality doesn't fit into any rating scheme or formula that we know of, and it can't quite make up for lack of technical knowledge. But without it, anyone's teaching value is probably low, and surely dimensionless.

September 1982

The Moral of the Tail

My grandmother used to warn me that the devil can cite scripture to his use, and I thought that was pretty neat. There is a similar problem concerning the canons of science. They are frequently misappropriated by folklore and then used for the justification or explanation of all sorts of silly things. It is well known that opposites attract, whether they be opposite electric charges or opposite temperaments. What goes up must come down, including baseballs, interest rates, and Roman empires. On the other hand, since Einstein, everything is relative, whether it be physical laws in different reference frames or moral codes in different cultures. When Darwin discovered the origin of the species, the apologists for laissez faire discovered social Darwinism. Evolution itself explained the success of the successful, and the law of averages showed that it would be prudent to take up the white man's burden. Of course, since Heisenberg restored free will, we cannot be certain of anything anymore. However, we can explain the housewife's burden in terms of the second law of thermodynamics. Everyone knows that it takes energy at the end of the day to bring order out of chaos.

Which brings me, quite naturally, to distribution curves. While working out a lecture for an introductory physics class, I was struck by the fact that at room temperature I am relatively stable. Yet at 586 K, at only twice room temperature, my goose and I would be cooked. Why should a factor of 2 in average molecular energy make such a difference in the rate of a chemical reaction? Why should a three-minute boiled egg take three minutes at 373 K, but at 363 K—don't wait! The chemists have a rule of thumb that the rate of a chemical interaction doubles for every ten degree rise in temperature, but that applies only to situations where the reactants are well above the threshold temperature for rapid consummation. There must be an energy barrier which two reactants must surmount before they can combine, or an energy minimum a molecule must absorb in order to disintegrate. The nature of the threshold may depend on whether the reaction absorbs energy or ends up releasing it. But in either case, the reaction ought to go with a rush as soon as the kinetic energy of the participants is slightly greater than the threshold.

Of course, it doesn't work like that. At 374 K, the egg still takes almost three minutes.

Molecules, like people, do not all have the same energy. At a given temperature there is a distribution of kinetic energies described by the Maxwell-Boltzmann formula:

$$n(E)dE = \frac{2\pi}{(\pi kT)^{3/2}} \, Ne^{-E/kT} \, \sqrt{E} \, dE.$$

The ordinate, $n(E)$, represents the density of molecules with a particular energy—the number per unit energy interval. The area under the curve of $n(E)$ versus E represents the total number of molecules, N. The scale factor for the energy unit is kT. Note that at $E = 0$, the exponential is equal to 1, but the density of molecules is 0 because of the factor of $\sqrt{E}$ in the distribution. Evidently, because of all the vibrating and milling around among the molecules, it's extremely unlikely to find a molecule just sitting still. At energies slightly above 0, the $\sqrt{E}$ term dominates, yielding an increase in the density of molecules having a particular energy. Soon, however, the negative exponential takes over, and the density at higher energies decreases steadily and rapidly. Before long we are left with only a negligible number of high energy molecules in the tail. But out there is where the action is.

Let's put numbers in, starting out with the size of the energy scale factor, kT. In terms of joules, the Boltzmann constant is: $k = 1.38 \times 10^{-23}$ J/K. The chemists prefer to discuss these details in terms of the universal gas constant ($R = N_0 k = 1.987$ kcal/kmol·K), but we, more sensibly, will use k in terms of electron volts: $k = 8.63 \times 10^{-5}$ ev/K. At room temperature of 293 K, $kT = 2.53 \times 10^{-2}$ eV $\cong$ 1/40 ev. The *average* kinetic energy of this distribution turns out to be the same as the average kinetic energy of a monatomic gas molecule at this temperature: $E = 3/2 \, kT$, which is about 1/25 ev. In the chemists' terminology, that's about 900 kcal/kmol. No wonder we're stable at room temperature! The binding energies of most molecules are of the order of 1 ev—or 25,000 kcal/kmol. If the average collision energy of an unstable molecule in our body is only 1/25 of the threshold energy for chemical disintegration, we won't decay until doomsday. Unless, perhaps, we're doomed by those few energetic molecules in the tail of the distribution. Let's see how many there are at threshold energy or beyond.

If $E_{\text{threshold}} = 1$ ev, then the fraction in the tail with energies greater than the threshold is:

$$\frac{N(> E_t)}{N} = \frac{2\pi}{(\pi kT)^{3/2}} \int_{E_t}^{\infty} e^{-E/kT} \sqrt{E}\, dE.$$

This integral cannot be reduced in closed form, but for our present purposes it is a good approximation that $\sqrt{E}$, which is a slowly varying function compared with the exponential, remains constant at the value, E_t. The fraction of molecules with enough energy to participate in the reaction thus becomes:

$$N(> E_t)/N \cong \sqrt{(E_t/kT)}\, e^{-(E_t/kT)}.$$

At room temperature, where $kT \cong 1/40$ ev, the value of this crucial fraction is:

$$\sqrt{40}\, e^{-40} \cong 3 \times 10^{-17}.$$

The small size of the crucial fraction is comforting unless we remember that there are 6×10^{23} molecules in every mole, and that means that in every mole of us there are 2×10^7 molecules with energies over the threshold. That may explain why we feel so dissipated in the summertime, but how can we explain the disastrous increase if we double the temperature to 586 K? After all, the average kinetic energy in the distribution will then be only 0.0833 ev, still a safe factor of 12 below the threshold. But look what has happened to the population of the tail. The crucial fraction above the threshold is now:

$$\sqrt{20}\, e^{-20} = 1 \times 10^{-8}.$$

A factor of 2 in temperature (or 2 in the threshold energy) changes the population of the tail by a factor of 3×10^8. As those molecules above the threshold combine or decay, their places are taken by other molecules from the main body of the distribution. You're cooked.

Is there a moral to this tale? Probably not. It would be too tempting for physics teachers to claim that in the distribution of human talents the crucial accomplishments take place at levels above a certain threshold. If we were to bemuse ourselves with such an elitist thought, we should also be constrained to remember that a small increase in the average value of a population yields enormous dividends in the size of the crucial fraction.

Not everything that goes up comes down these days. Some quantities remain invariant regardless of the reference frame. And, fortunately, the distribution of human talents and the elusive nature of accomplishment are far too complex to be pinned down and analyzed in the tail of a graph.

October 1977

Physics, Where the Least Action Is

hristmas is a time for mysteries, both profound and trivial. Our annual physics poster invokes a principle that may be one or the other. *Physics, Where the least action is?* Certainly! Where else but in physics is there informed concern about implications of the great conservation laws, symmetries, and variational principles? By *least action,* of course, we mean the principle proposed by Maupertuis, and given its modern formulation by Lagrange and Hamilton. We also mean Fermat's principle of least time.

Over two thousand years ago Hero of Alexandria (the same Hero who invented a rotary steam engine) proved mathematically that a light ray reflected from a plane mirror takes a path shorter than any other path. (Only paths leading from source to detector by way of the mirror are considered.) This classic and clever proof is given in many of our modern introductory physics texts. Here is a sketch of the geometry involved. The requirement of minimum path automatically yields the equality of incident and reflected angles.

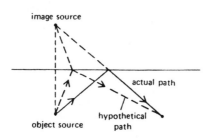

In the middle of the seventeenth century, Fermat extended the principle to light rays passing through various media, such as glass and air. Here the rule is transformed from "shortest path" to "least time," raising the possibility of some grand economy of nature. Regardless of the philosophical implications, Fermat's principle and its corollaries can serve as practical guides in the construction of optical instruments. To find a focusing geometry, instead of applying Snell's law concerning incident and refracted angles at each interface, simply look for surfaces that will bring all rays to a point with the same travel times. For instance, when parallel light passes through a focusing lens, the center ray clearly travels a shorter *distance* to the focus than does the fringe ray. The center ray, however, travels for a while at reduced speed through a thicker piece of glass. If the lens focuses, both rays take the same time.

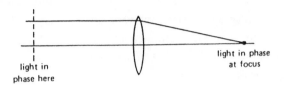

light in phase here

light in phase at focus

Nature really doesn't always economize in these matters. All that is guaranteed by Fermat's principle is that the time taken by light to go from one point to another is an *extremum*. Usually it is a minimum but it can be a maximum.

The comparison of times should always be made between paths that begin together at a point and end together at some other point and differ only slightly in midcourse. (For instance, times for light rays should be compared only if they bounce off the same mirror.) To see an example where the natural path of light is a *maximum*, sketch and compare several paths of light from a point to a concave parabolic mirror and back to another point.

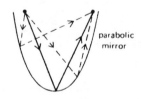

parabolic mirror

In 1744 the French mathematician Maupertuis announced his discovery of the principle of least action. He claimed that mechanical processes always occur in such a way that the product of mass, velocity, and distance (properly interpreted) is a minimum. In modern terminology, if the product of momentum and displacement is summed for all elements of path as an object goes from one point in space to another, and if energy is conserved, then this sum is a minimum or maximum for the natural path compared with any other path. To be mathematically precise:

$$\delta \int_{s_1}^{s_2} p\,ds = 0,$$

where δ means a small variation in path for the integral. A more complete and useful variational principle makes use of the Lagrangian function. In classical mechanics this is simply the difference between the kinetic and potential energy: $\mathcal{L} = E_{kin} - E_{pot}$. Hamilton expressed the "least action" variational principle as:

$$\delta \int_{t_1}^{t_2} \mathcal{L}\,dt = 0.$$

Conservation of energy is not assumed in this formulation; it tumbles out of the equation along with Newton's laws (in the classical case).

All the different paths for the integral must start at the same point in space and time and must end at another particular point, taking identical times. The principle is that the integral taken along the true path must be one where slight changes in path will not cause a first order variation in the integral. This is a normal feature of a minimum or maximum. A small departure of the variable from the equilibrium point makes only a second order change in the function. For instance, at the bottom of the parabola, $y = x^2$, if x varies by Δx, y varies by $2x\Delta x + \Delta x^2$. Since at the bottom (minimum point) $x = 0$, then $\Delta y = 0 + \Delta x^2$, a second-order effect.

In Feynman's *Lectures* (Vol. II, 19-1), he gives some simple examples of the application of the least-action principle. For instance, for an object moving horizontally in the gravitational field, the potential energy term in the Lagrangian can be ignored since there will be no variation in it for any horizontal path. What remains is the condition:

$$\delta \int_{t_1}^{t_2} E_{kin}\,dt = \delta \int_{t_1}^{t_2} \tfrac{1}{2}\,mv^2 dt = 0.$$

The integration path (as opposed to the horizontal path of the object) could be one of constant velocity or it could be one where the object first had large constant velocity and then small.

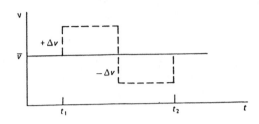

The only requirement is that the total time $(t_2 - t_1)$ be the same for all possible types of trips over the same distance. Therefore, the average speed must be the same. Suppose the object travels at higher than average speed for the first half of the time, and then, necessarily slower for the second half. The integral becomes a sum of two parts:

$$\int_{t_1}^{t_2} \mathscr{L} \, dt = \tfrac{1}{2}m \left[(\bar{v} + \Delta v)^2 \, \frac{\Delta t}{2} + (\bar{v} - \Delta v)^2 \, \frac{\Delta t}{2} \right] = \tfrac{1}{2} \, m\bar{v}^2 \, \Delta t + \tfrac{1}{2} \, m(\Delta v)^2 \, \Delta t$$

This is larger than the integral when the velocity remains constant, and because of the square term, any variation of velocities greater or less than the average will produce an integral with a larger value. In this case the variational principle requires constant velocity. Note that energy is not conserved for paths other than the true one.

If an object is thrown straight up, the required motion involves potential energy. Constant velocity will no longer yield the smallest integral. To reduce the value of the integral, the object should rapidly move to a region where the potential energy is large, thus making $(E_{kin} - E_{pot})$ small. The solution to the general problem yields Newton's second law, from which the parabolic nature of this particular path can be calculated.

In the diagrams below are descriptions of two different paths for the flight of an object straight up and down. For the flight that actually takes place, the velocity linearly decreases from maximum positive (up) to maximum negative (down). The kinetic energy, potential energy, and

Lagrangian are shown plotted as a function of time. The integral of least action corresponds to the area under the Lagrangian curve. The two positive regions cancel out only a small fraction of the negative area, leaving a large negative value. An alternative path might be one where the object climbs up with constant velocity to the same height as in the true path. The v (t) graph is shown for this case. In the plot of E_{kin}, E_{pot}, and $\mathcal{L}$ for this path, it is clear that the area under the $\mathcal{L}$ curve is smaller *negative* than that for the true path. Therefore, the least action integral is less (more negative) for the true path.

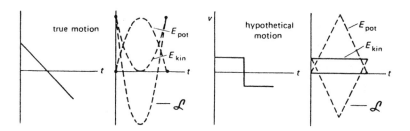

Whether or not a variational principle exists for a particular phenomenon must be determined in the final analysis by whether or not the laws of motion (or equivalently for light, the law of refraction) can be deduced from the principle.

Sometimes calculations are easier to do starting with the integral form; more often they are easier with the laws of motion in the more widely known differential form. There is, however, a major philosophical difference between the two forms. In the differential form (e.g., $F = ma$) if an object changes velocity at a point it is because there is a force acting on it at that point. In the integral form, it appears as if the object must somehow know which of all possible paths will produce the least (or maximum) action. In the case of light, perhaps the wavefronts probe various routes, calculate the times, and then choose the extremum. In anthropomorphic terms, it sounds as if Nature uses the end to choose the means.

It isn't like that of course. The variational principles are clever for concise formulations, fruitful in suggesting analogies, and sometimes powerful in calculations. Always, however, their validity hinges on the deduced laws of motion. In the case of wave motion the choice of optimum paths (whether minimum or maximum) can be interpreted in terms of phase interferences.

Nevertheless, there are mysteries in this business. Indeed, the primary concerns of physicists are the primeval mysteries. One of these is how man has learned to reduce and summarize vast realms of experience into a few powerful generalizations. Where else but in physics will you find the least action?

December 1972

DATE DUE

GAYLORD   #3522PI   Printed in USA